Molecular Breeding of Forage and Turf

Hikmet Budak • German Spangenberg
Editors

Molecular Breeding of Forage and Turf

The Proceedings of the 8th International Symposium on the Molecular Breeding of Forage and Turf

Springer

Editors
Hikmet Budak
Molecular Biology, Genetics and Bioengineering, Faculty of Engineering and Natural Sciences
Sabanci University
Istanbul
Turkey

German Spangenberg
Department of Environment and Primary Industries
La Trobe University
Bundoora, Victoria
Australia

ISBN 978-3-319-08713-9 ISBN 978-3-319-08714-6 (eBook)
DOI 10.1007/978-3-319-08714-6

Library of Congress Control Number: 2015930836

Springer Cham Heidelberg New York Dordrecht London

Printed on acid-free paper

Springer is part of Springer Science+Business Media (www.springer.com)

Preface

Grasslands in the form of both forage and turf are the life support of the planet providing sustenance to wildlife, livestock, and thus to humans. They anchor the soil preventing erosion, harnessing the freshwater resources, and creating an environment for outdoor sports, recreation, and entertainment. They also serve to beautify our environment along with their physical and nutritional value. Bioenergy is also obtained from the grasslands and they have recently become of all the more importance for providing biofuel alongside food. In recent years however this useful resource has been afflicted with environmental insults such as increasing drought spells globally and it is important to further manipulate this resilient and versatile resource for human benefit.

Previously, the 6th MBFT Symposium was held in Buenos Aires, Argentina and the 7th MBFT Symposium was held in Salt Lake City, USA. The 8th International Symposium on Molecular Breeding of Forage and Turf was held in Sabanci University, Istanbul, Turkey. From amongst the attendees there were scientists from 15 countries from all fields of plant biology including geneticists, and molecular biologists as well as breeders and agronomists. The meeting encompassed oral presentations from leading scientists on molecular plant breeding, a surmounting number of diverse poster presentations and also tours of the historic Old Istanbul City and a boat tour of the Bosphorus Strait.

This book features papers from oral presentations of the symposium. It extensively covers the various themes discussed along with definitive reports shedding light on recent developments in systems biology, functional genomics, and application of molecular breeding in forage and turf.

The 8th MBFT Symposium and the publication of all its proceedings in this book, *Molecular Breeding of Forage and Turf*, have been supported by the Bioengineering and Biological Sciences Program, Faculty of Engineering and Natural Sciences, Sabanci University; the Samuel Roberts Noble Foundation, and the Scientific and the Technological Research Council of Turkey.

We thank once again the International and Local Organizing Committees for their efforts in making the symposium such a great success.

Lastly we also thank our authors for their relentless effort and work contributing to the publication of the *Molecular Breeding of Forage and Turf.*

September 2014

Prof. Dr. Hikmet Budak
Prof. Dr. German Spangenberg

Contents

Contributors

Bala Ani Akpinar Genetics and Bioengineering Program, Molecular Biology, Engineering and Natural Sciences, Sabanci University, 34956 Tuzla, Istanbul, Turkey

B. Alarcon-Zuniga Department of Animal Science, Universidad Autonoma Chapingo, Km 38.5. Hwg Mexico-Texcoco, Chapingo, Mex 56230, Mexico

A. Aleliūnas Institute of Agriculture, Lithuanian Research Centre for Agriculture and Forestry, Instituto a. 1, Akademija, 8344 Kėdainiai distr, Lithuania

Kazunori Ashikaga National Agricultural and Food Research Organization, Hokkaido Agricultural Research Center, 1 Hitsujigaoka, Toyohira, Sapporo 062-8555, Hokkaido, Japan

T. Asp Molecular Biology and Genetics, Aarhus University, Forsøgsvej 1, 4200 Slagelse, Denmark

P. Azhaguvel Syngenta, 2369–330th Street, Slater, IA 50244, USA

Aike Bao State Key Laboratory of Grassland Agro-ecosystems, College of Pastoral Agriculture Science and Technology, Lanzhou University, 730020 Lanzhou, P. R. China

A. R. Blount North Florida Research & Education Center, Institute of Food and Agricultural Sciences, University of Florida, Mariana, FL 32446, USA

G. Brazauskas Institute of Agriculture, Lithuanian Research Centre for Agriculture and Forestry, Instituto a. 1, Akademija, 8344 Kėdainiai distr, Lithuania

Hikmet Budak Molecular Biology, Genetics and Bioengineering, Faculty of Engineering and Natural Sciences, Sabanci University, 34956 Tuzla, Istanbul, Turkey

S. Byrne Molecular Biology and Genetics, Aarhus University, Forsøgsvej 1, 4200 Slagelse, Denmark

Ozge Canli Genetics and Bioengineering Program, Molecular Biology, Engineering and Natural Sciences, Sabanci University, 34956 Tuzla, Istanbul, Turkey

T. Cervantes-Martinez Department of Animal Science, Universidad Autonoma Chapingo, Km 38.5. Hwg Mexico-Texcoco, Chapingo, Mex 56230, Mexico

K. Chekhovskiy Forage Improvement Division, The Samuel Roberts Noble Foundation, 2510 Sam Noble Parkway, Ardmore, OK 73401, USA

Shuangyan Chen Key Laboratory of Plant Resources, Institute of Botany, Chinese Academy of Sciences, Beijing, P. R. China

Lindsay V. Clark Department of Crop Sciences, University of Illinois, Urbana-Champaign, Urbana, IL 61801, USA

Klaus J. Dehmer Leibniz Institute of Plant Genetics and Crop Plant Research, Corrensstrasse 3, Gatersleben 06466, Germany

Jiangli Dong State Key Laboratory of Agrobiotechnology, College of Biological Sciences, China Agricultural University, 100193 Beijing, P. R. China

J. C. Dubeux Jr. North Florida Research & Education Center, Institute of Food and Agricultural Sciences, University of Florida, Mariana, FL 32446, USA

Maria S. Dwiyanti Field Science Center for Northern Biosphere, Hokkaido University, Kita 11 Nishi 10, Sapporo, Hokkaido 060–0811, Japan

N. W. Ellison AgResearch Grasslands, Private Bag 11008, Palmerston North, New Zealand

J. A. Ferrell Agronomy Department, Institute of Food and Agricultural Sciences, University of Florida, Gainesville, FL 32611, USA

Hiroki Fujii Hokkaido Research Organization, Kitami Agricultural Experiment Station, 52 Yayoi, Kunneppu, Tokoro-gun, , Sapporo 099-1496, Hokkaido, Japan

Qing Ge-Le Inner Mongolia Academy of Agricultural & Animal Husbandry Sciences, Hohhot 010031, P. R. China

Andreas Graner Leibniz Institute of Plant Genetics and Crop Plant Research, Corrensstrasse 3, Gatersleben 06466, Germany

A. Haegeman Plant Sciences Unit—Growth and Development, Institute for Agricultural and Fisheries Research (ILVO), Caritasstraat 21, 9090 Melle, Belgium

Yoichiro Hoshino Field Science Center for Northern Biosphere, Hokkaido University, Sapporo, Japan

Linkai Huang Department of Grassland Science, Animal Science and Technology College, Sichuan Agricultural University, Ya'an, P. R. China

Louise Bach Jensen Illumina Denmark ApS, Ørestads Boulevard 73, 2300, København, Denmark

Yi Jing Inner Mongolia Academy of Agricultural & Animal Husbandry Sciences, Hohhot 010031, P. R. China

K. Jonavičienė Institute of Agriculture, Lithuanian Research Centre for Agriculture and Forestry, Instituto a. 1, Akademija, 8344 Kėdainiai distr, Lithuania

Mohsen Kafi Department of Horticulture Science, Faculty of Agriculture sciences and engineering, College of Agriculture & Natural Resources, University of Tehran, Karaj, Iran

V. Kemešytė Institute of Agriculture, Lithuanian Research Centre for Agriculture and Forestry, Instituto a. 1, Akademija, 8344 Kėdainiai distr, Lithuani

Nayer Azam Khoshkholghsima Agriculture Biotechnology Research Institute of Iran (ABRII), P.O. Box 31535-1897, Karaj, Iran

Gongshe Liu Key Laboratory of Plant Resources, Institute of Botany, Chinese Academy of Sciences, Beijing, P. R. China

Hongkui Liu Inner Mongolia Academy of Agricultural and Animal Husbandry Sciences, Hohhot 010031, P. R. China

Yong-Zhi Liu Inner Mongolia Academy of Agricultural & Animal Husbandry Sciences, Hohhot 010031, P. R. China

Thomas Lübberstedt Agronomy Department, Iowa State University, 1204 Agronomy, Ames, IA 50011, USA

J. W. Miles International Center for Tropical Agriculture (CIAT), A.A. 6713, Cali, Colombia

S. Mukhergee Computing Services, The Samuel Roberts Noble Foundation, 2510 Sam Noble Parkway, Ardmore, OK 73401, USA

P. R. Munoz Agronomy Department, Institute of Food and Agricultural Sciences, University of Florida, Gainesville, FL 32611, USA

H. Muylle Plant Sciences Unit—Growth and Development, Institute for Agricultural and Fisheries Research (ILVO), Caritasstraat 21, 9090 Melle, Belgium

Hironori Nagano Field Science Center for Northern Biosphere, Hokkaido University, Kita 11 Nishi 10, Sapporo, Hokkaido 060–0811, Japan

Lizhen Nie Inner Mongolia Academy of Agricultural & Animal Husbandry Sciences, Hohhot 010031, P. R. China

Yongzhen Pang Key Laboratory of Plant Resources, Institute of Botany, Chinese Academy of Sciences, Beijing, P. R. China

K. H. Quesenberry Agronomy Department, Institute of Food and Agricultural Sciences, University of Florida, Gainesville, FL 32611, USA

S. Quijano-Hernandez Department of Animal Science, Universidad Autonoma Chapingo, Km 38.5. Hwg Mexico-Texcoco, Chapingo, Mex 56230, Mexico

J. J. M. Ramirez-Gonzalez Instituto Nacional de Investigaciones Forestales Agrícolas y Pecuarias, Experimental Field Station—Golfo Centro, Hueytamalco, Pue 73588, Mexico

Iman Rohollahi Department of Horticulture Science, Faculty of Agriculture sciences and engineering, College of Agriculture & Natural Resources, University of Tehran, Karaj, Iran

I. Roldán-Ruiz Plant Sciences Unit—Growth and Development, Institute for Agricultural and Fisheries Research (ILVO), Caritasstraat 21, 9090 Melle, Belgium

T. Ruttink Plant Sciences Unit—Growth and Development, Institute for Agricultural and Fisheries Research (ILVO), Caritasstraat 21, 9090 Melle, Belgium

Erik J. Sacks Department of Crop Sciences, University of Illinois, Urbana-Champaign, Urbana, IL 61801, USA

Malay C. Saha Forage Improvement Division, The Samuel Roberts Noble Foundation, 2510 Sam Noble Parkway, Ardmore, OK 73401, USA

Uwe Scholz Leibniz Institute of Plant Genetics and Crop Plant Research, Corrensstrasse 3, Gatersleben 06466, Germany

Tatjana Sretenovic Rajicic Leibniz Institute of Plant Genetics and Crop Plant Research, Corrensstrasse 3, Gatersleben 06466, Germany

G. Statkevičiūtė Institute of Agriculture, Lithuanian Research Centre for Agriculture and Forestry, Instituto a. 1, Akademija, 8344 Kėdainiai distr, Lithuania

A. V. Stewart PGG Wrightson Seeds, PO Box 175, Lincoln 7640, New Zealand

Jie Sun Inner Mongolia Academy of Agricultural & Animal Husbandry Sciences, Hohhot 010031, P. R. China

Zhan-Min Sun Biotechnology Research Institute, Chinese Academy of Agricultural Sciences, Beijing 100081, P. R. China Tel.: 00861082106113 e-mail: wuyanmin211@gmail.com

State Key Laboratory of Plant Physiology and Biochemistry, College of Biological Sciences, China Agricultural University, Beijing 100094, P. R. China

S. K. Talukder Forage Improvement Division, The Samuel Roberts Noble Foundation, 2510 Sam Noble Parkway, Ardmore, OK 73401, USA

Hiroyuki Tamaki National Institute of Livestock and Grassland Science, 768 Senbonmatsu, Nasusiobara, Tochigi 329-2793, Japan

Ken-ichi Tamura National Agricultural and Food Research Organization, Hokkaido Agricultural Research Center, 1 Hitsujigaoka, Toyohira, Sapporo 062-8555, Hokkaido, Japan

Tsuneki Tanaka Hokkaido Research Organization, Kitami Agricultural Experiment Station, 52 Yayoi, Kunneppu, Tokoro-gun, , Sapporo 099-1496, Hokkaido, Japan

Yi-Xiong Tang Biotechnology Research Institute, Chinese Academy of Agricultural Sciences, Beijing 100081, P. R. China

S. Van Glabeke Plant Sciences Unit—Growth and Development, Institute for Agricultural and Fisheries Research (ILVO), Caritasstraat 21, 9090 Melle, Belgium

F. van Parijs Plant Sciences Unit—Growth and Development, Institute for Agricultural and Fisheries Research (ILVO), Caritasstraat 21, 9090 Melle, Belgium

Suomin Wang State Key Laboratory of Grassland Agro-ecosystems, College of Pastoral Agriculture Science and Technology, Lanzhou University, 730020 Lanzhou, P. R. China

Tao Wang State Key Laboratory of Agrobiotechnology, College of Biological Sciences, China Agricultural University, 100193 Beijing, P. R. China

W. Eberhard Weber Institute of Agricultural and Nutritional Sciences, Plant Breeding, Martin-Luther-University Halle-Wittenberg, Halle 06099, Germany

M. L. Worthington International Center for Tropical Agriculture (CIAT), A.A. 6713, Cali, Colombia

Yanmin Wu Biotechnology Research Institute, Chinese Academy of Agricultural Sciences, 100081 Beijing, P. R. China

Xing-Guo Xiao State Key Laboratory of Plant Physiology and Biochemistry, College of Biological Sciences, China Agricultural University, Beijing 100094, P. R. China

Toshihiko Yamada Field Science Center for Northern Biosphere, Hokkaido University, Kita 11 Nishi 10, Sapporo 060-0811, Hokkaido, Japan

Qiong-Lin Zhang Inner Mongolia Academy of Agricultural & Animal Husbandry Sciences, Ltd., Hohhot 010030, P. R. China

Xinquan Zhang Department of Grassland Science, Animal Science and Technology College, Sichuan Agricultural University, Ya'an, P. R. China

Heping Zhao Inner Mongolia Academy of Agricultural & Animal Husbandry Sciences, Hohhot 010031, P. R. China

Mei-Lang Zhou Biotechnology Research Institute, Chinese Academy of Agricultural Sciences, 100081 Beijing, P. R. China

Yi Xiong [illegible]
[illegible] Science [illegible]

S. [illegible]
[illegible]

[illegible]
[illegible]

[illegible] Ma[illegible]

Chapter 1
Deciphering Drought Tolerance in Tall Fescue [*Lolium arundinaceum* (Schreb.) Darbysh.]

Malay C. Saha, S.K. Talukder, P. Azhaguvel, S. Mukhergee and K. Chekhovskiy

1.1 Introduction

Drought is the single most important constraint to crop productivity causing yield loss up to 50 % or more (Boyer 1982). Recently, severe droughts have been prevailed in the Southern Great Plains of the USA. During the unprecedented drought in 2011, Oklahoma experienced driest 4 months since 1921 and accounted for $ 1.6 billion losses in the drought-related agriculture. Agricultural losses in Texas were estimated $ 7.62 billion that made 2011 the costliest drought to date. Tall fescue [*Lolium arundinaceum* (Schreb.) Darbysh.] is an important cool season hay and pasture grass grown in over 14 million ha in the USA (Sleper and West 1996). The persistence of this perennial grass largely depends on their ability to tolerate drought stress. Tall fescue cultivars do not generally persist for more than 2–4 years in the south central USA due to drought stress (Hopkins 2005). Thus, developing drought-tolerant cultivar is the key strategy for improving productivity and persistence of the Continental tall fescue in the region.

Drought tolerance is difficult to select for because of low heritability occasioned by nonuniform testing conditions and large genotype-by-environment interactions. Simple, reliable, and repeatable measures of drought tolerance can facilitate rapid screening of large number of genotypes to identify superior types. Relative water content (RWC) is a measure of plant water status in terms of physiological cellular water deficit (Barrs and Weatherly 1962). Increased RWC under water deficit is

M. C. Saha (✉) · S. Talukder · P. Azhaguvel · K. Chekhovskiy
Forage Improvement Division, The Samuel Roberts Noble Foundation,
2510 Sam Noble Parkway, Ardmore, OK 73401, USA
e-mail: mcsaha@noble.org

P. Azhaguvel
Syngenta, 2369–330th Street, Slater, IA 50244, USA

S. Mukhergee
Computing Services, The Samuel Roberts Noble Foundation, 2510 Sam Noble Parkway, Ardmore, OK 73401, USA

H. Budak, G. Spangenberg (eds.), *Molecular Breeding of Forage and Turf*,
DOI 10.1007/978-3-319-08714-6_1

associated with the increased drought tolerance. Increased osmotic potential (OP) under water stress has been linked to tissue survival and regrowth. Both RWC and OP could be used as assays to identify drought-tolerant tall fescue genotypes from natural populations (Elmi and West 1995).

Various physiological, metabolic, and defense mechanisms are activated during drought stresses, which make a plant to survive and/or maintain growth and reproduction (Valliyodan and Nguyen 2006). Though various genes and signal transduction involved with drought tolerance have been studied in different species (Chinnusamy et al. 2004; Li et al. 2012), no information is available in tall fescue. Currently, there have been only 63,853 tall fescue ESTs available in the gene bank database. Thus, the objectives of our projects are to: (i) develop a rapid and reliable drought screening protocol for tall fescue germplasms, (ii) genotype and phenotype a mapping population to identify QTL associated with important drought-related traits, and (iii) identify drought-tolerant key genes and genetic factors in tall fescue through transcriptome profiling.

1.2 Materials and Methods

A total of 1000 genotypes of the Noble Foundation Continental tall fescue breeding population PDF584 (released as "Texoma MaxQ II") were evaluated under well watered and drought stressed conditions in the Foundation's greenhouse. These plants were clonally propagated into two replicates. The temperature was set at 24 °C and the photoperiod was maintained 16 h. For each genotype, three fully developed collared leaves were collected from three tillers in each replicate. Data for RWC and OP were gathered for all samples. RWC was determined using 2-cm long-leaf samples from each of the genotypes and replicates (Barrs and Weatherly 1962). RWC was calculated as: RWC=(FW-DW)/(TW-DW)*100. The OP was determined using a vapor pressure osmometer (Wescor's VAPRO Vapor pressure osmometer 5520, Westcor, Logan, Utah). The osmolality units (mmol kg^{-1}) were converted to MPa using the Van't Hoff relation; $\psi_S = -CiRT$, where C is the osmolality value in mol kg^{-1}, i is an ionizing constant assumed equal to unity; R is the ideal gas constant (0.0083143 kg MPa $mol^{-1}K^{-1}$), and T is absolute temperature (K=°C+273) (Nobel 1983).

Tall fescue genotypes, B400 with low OP and high RWC during drought and W279 with high OP and low RWC were grown in growth chamber. Pots were vernalized for 40 days at 4 °C followed by 15 days of acclimation under optimum light (10 h), temperature (24 °C), and water (frequently irrigated) conditions. Plants were then subjected under water stress for 15 days by providing minimum amount of water (stopped watering till wilting sign appeared and watered again just for the survivability of the plants). Samples of leaf, stem, root, and inflorescence from all the plants were collected. RNA was isolated using TriReagent and purified with RNeasy Plant Mini Kit (Qiagen, Valencia, CA). RNA quality was substantiated by 2100 Bioanalyzer RNA Nanochip (Agilent, Santa Clara, CA) and quantified using NanoDrop ND-1000 spectrophotometer (NanoDrop, Wilmington, DE). A total of 20 μg RNA was pooled from the four tissue types of three replications for cDNA library preparation. Double-stranded cDNA was obtained using SuperScript Double-Stranded cDNA Synthesis kit (Invitrogen, Camarillo, CA). Sequencing was performed at the National Center for Genome Resources (NCGR), Santa Fe, New Mexico, USA using Illumina Genome Analyzer IIx system.

Greenhouse data analysis was performed using the PROC MIXED procedure of SAS. The sequenced reads were processed with a custom R script based on short-read package (Morgan et al. 2009). The high quality filtered reads were then assembled by Trinityrnaseq_r2013-02-25 with strand specific option "–SS_lib_type" set to "F" and "min_kmer_cov" set to 2 (Grabherr et al. 2011). The resulting contigs were clustered and further assembled to reference transcripts by Trinityrnaseq_r2013-02-25. Read mapping and quantification of the reference transcripts were done using Tuxedo suite (Tophat, Bowtie, Cufflinks). Assembled transcriptome data were evaluated by using blastn in tall fescue EST database and *Brachypodium* genome database. All the sequences were used for blast search against our transcriptome database using an E-value threshold of e^{-5} (0.00001). The search for sequence homology was performed by using BLAST v2.2.25+(http://blast.ncbi.nlm.nih.gov/Blast.cgi) with an E-value cut-off of e^{-5} (0.00001). SSRs were detected using MIcroSAtellite Identification Tool (MISA v1.0).

1.3 Results and Discussions

The 1000 tall fescue genotypes differed widely for RWC (range 33.7–97.3 %, mean 79.7 %), OP (range −0.5–−2.4 MPa, mean −1.2 MPa) and chlorophyll content (range 25.8–62.8, mean 41.0). Only nine genotypes were identified which had RWC, OP, and chlorophyll content 10 % above the mean (Fig. 1.1). There were

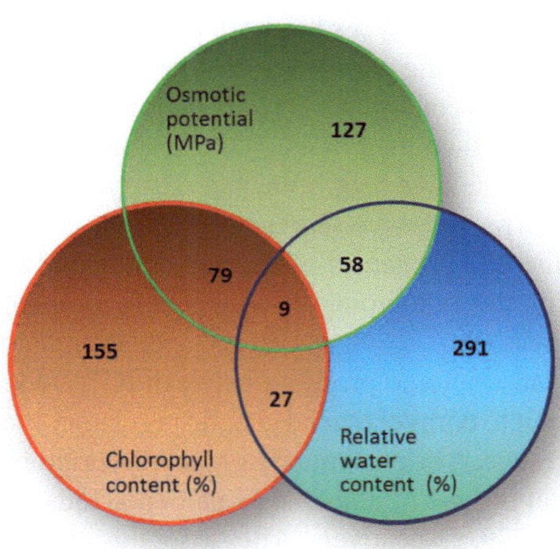

Fig. 1.1 A Venn diagram showing the number of tall fescue genotypes with mean scores 10 % above the mean of the population

Table 1.1 Tall fescue genotypes with relative water content (RWC) and osmotic potential (OP) below or above the LSD units when evaluated in greenhouse experiments

Trait	LSD units below			LSD units above			
	−2	−1	0	0	1	2	3
OP	16	168	357	273	152	27	6
RWC	2	136	310	471	80	0	0

27, 58, and 79 genotypes which had chlorophyll content and RWC, RWC and OP, and OP and chlorophyll content 10% above the mean, respectively. RWC has been proposed as the most integrative measure of drought tolerance (Blum 1999). Osmotic adjustment is an important process of plant adaptation to drought because it conserves cellular hydration under stress. Increased OP under water stress has been linked to tissue survival and regrowth (Elmi and West 1995). Both RWC and OP could be used as assays to identify genotypes with high drought tolerance. Most of the genotypes had RWC and OP within one LSD unit of the mean, but some genotypes had OP values over two LSD units above and below the mean. Only two genotypes had RWC values below two LSD units (Table 1.1). There were six genotypes with OP values above three LSD units. The results indicated that the genotypes we evaluated had different physiologies and differed widely for RWC and OP.

The RWC and OP were negatively correlated ($r=-0.68$, $P<0.0001$) for all genotypes. This implicates that overall observed changes in RWC might be influenced by differences in OP, but not necessarily by cell water volume change. Based on RWC and OP, the 25 most tolerant and 25 most susceptible genotypes were selected for further evaluations. After two sets of field and greenhouse experiments, we finally selected the most contrasting genotypes for drought tolerance. Genotypes NFTD348 and NFTD400 were identified the most tolerant, while NFTD279 and NFTD947 were identified as the most susceptible in the population. Across all studies, NFTD400 and NFTD348 were 85.7 and 76.2% times above the LSmeans, while genotypes NFTD279 and NFTD947 were 81.0 and 85.7% times below the mean, respectively. NFTD400 and NFTD279 were crossed and a mapping population of 252 genotypes has been developed.

The mapping population has been evaluated in field experiments under irrigated and rainfed conditions. Data on various morphological and physiological traits, e.g., chlorophyll content, RWC, OP, heading date, plant height, recovery time, and biomass yield, have been collected. The two parents were very distinct for most of the traits. Transgressive segregation was observed in the population (data not shown). All the 252 progenies of the population and the parents have been genotyped following the genotyping-by-sequencing protocol. Mapping and QTL analyses are in progress.

Transcriptome profiling between NFTD400 and NFTD279 genotypes was carried out to unravel crucial genetic regulatory mechanism of water stress responses in tall fescue. A brief summary of the transcriptome assembly is presented in Table 1.2. A total of 39.4 M and 58.6 M pair end reads were obtained from the NFTD400 and NFTD279, respectively. After assembly, 199,399 reference transcripts were recovered with an average read length of 585 bp (Table 1.2). A total of 2986 transcripts

Table 1.2 Summary of transcriptome assembly data obtained from two tall fescue genotypes contrasting for drought tolerance characteristics

Item	Measurements
Number of paired end reads of B400	39395238*2
Number of paired end reads of W279	58648268*2
Total number of reference transcript	199,399
Maximum transcript length	9671
Minimum transcript length	201
Average transcript length	585
Total transcript length (bases)	116,688,388
Actual read length	54 × 2
Quality trimmed read length	33 × 2
N50	752
E-value cutoff	0.00001

were significantly differentially expressed between the two genotypes. Thousand of them were found to be annotated and associated with metabolic pathways and enzyme coding genes.

MIcroSAtellite Identification Tool (MISA v1.0) was used to identify SSRs in the transcriptomes. The minimum number of repeats used to report a dinucleotide SSR was eight, six for a tri-nucleotide repeat, and four for tetra-nucleotide and above. A total of 8788 SSRs were identified in 8490 sequences of which, 243 sequences contained multiple SSRs. Primer-3 software was used to call primer pairs (PPs) and a total of 6348 PPs were developed. Among these potential SSRs, 313 had di-nucleotide, 1207 had tri-nucleotide, 826 had tetra-nucleotide, 1727 penta-nucleotide, 2082 hexa-nucleotide repeat motifs, and rest 193 had compound repeats. On an average, one SSR was found in every 13.36 kb with a frequency of 4.38 %. All the SSR containing sequences were then aligned and mapped into *Brachypodium* chromosomes sequence using blastn. In general, the SSRs were evenly distributed throughout all five *Brachypodium* chromosomes (Fig. 1.2). As expected, majority of the SSRs were concentrated in the distal part of each chromosomes and the centromeric region was lacking any SSR. Similar distribution of EST-SSR markers were observed in tall fescue genetic map (Saha et al. 2005).

1.4 Conclusion

We are in the process to develop genetic and genomic resources for deciphering drought tolerance in tall fescue. Thousand genotypes of a tall fescue population were screened for chlorophyll content, RWC and OP. Most contrasting genotypes were selected through a series of greenhouse and field experiments. Two most contrasting genotypes were used to construct a mapping population. Genotyping and phenotyping of the population is in progress. Transcript profiling of the genotypes identified differentially expressed transcripts at drought-stressed conditions. We identified a set of SSR markers well distributed in the genome. All these resources

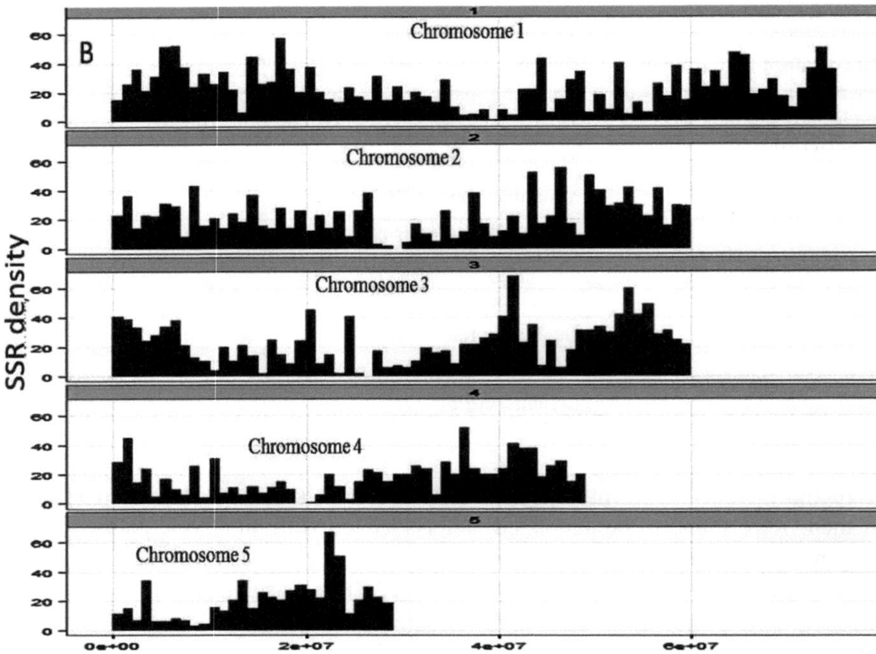

Fig. 1.2 Distribution of tall fescue EST-SSRs on *Brachypodium* chromosomes

can be used to expedite the cultivar development process in tall fescue, an outcrossing polyploid grass species.

References

Barrs HD, Weatherly PE (1962) A re-examination of the relative turgidity technique for estimating water deficit in leaves. Aust J Biol Sci 15:413–428

Blum A (1999) Towards standard assays of drought resistance in crop plants. pp 3–35. In: Proceedings of the Workshop on Molecular Approaches for the Genetic Improvement of Cereals for Stable Production in Water-limited Environments (Mexico: CIMMYT)

Boyer JS (1982) Plant productivity and environment. Science 218:443–448

Chinnusamy V, Schumaker K, Zhu JK (2004) Molecular genetic perspectives on cross-talk and specificity in abiotic stress signalling in plants. J Exp Bot 55:225–236

Elmi AA, West CP (1995) Endophyte infection effects on stomatal conductance, osmotic adjustment and drought recovery of tall fescue. New Phytol 31:61–67

Grabherr MG, Haas BJ, Yassour M, Levin JZ, Thompson DA, Amit I, et al (2011) Full-length transcriptome assembly from RNA-Seq data without a reference genome. Nat Biotechnol 29:644–652

Hopkins AA (2005) Grazing tolerance of cool-season grasses planted as seeded sward plots and spaced plants. Crop Sci 45:1559–1564

Li YC, Meng FR, Zhang CY, Zhang N, Sun MS, Ren JP, et al (2012) Comparative analysis of water stress-responsive transcriptomes in drought-susceptible and-tolerant wheat (*Triticum aestivum* L.). J Plant Biol 55:349–360

Morgan M, Anders S, Lawrence M, Aboyoun P, Pages H, Gentleman R (2009) ShortRead: a bioconductor package for input, quality assessment and exploration of high-throughput sequence data. Bioinformatics 25:2607–2608

Nobel PS (1983) Biophysical plant physiology and ecology. Freeman, San Francisco

Saha MC, Mian MAR, Zwonitzer JC, Chekhovskiy K, Hopkins AA (2005) An SSR- and AFLP-based genetic linkage map of tall fescue (*Festuca arundinacea* Schreb.). Theor Appl Genet 110:323–336

Sleper DA, West CP (1996) Tall fescue. In: Moser LE, Buxton DR, Casler MD (ed) Cool-season forage grasses. American Society of Agronomy; Crop Science Society of America; Soil Science Society of America, Madison, pp 471–502

Valliyodan B, Nguyen HT (2006) Understanding regulatory networks and engineering for enhanced drought tolerance in plants. Curr Opin Plant Biol 9:189–195

Chapter 2
Evaluation of Perennial Ryegrass Association Mapping Population for Freezing Tolerance Traits

A Aleliūnas, G Statkevičiūtė, K Jonavičienė, V Kemešytė and G Brazauskas

2.1 Introduction

Perennial ryegrass (*Lolium perenne* L.) is one of the most important agricultural cool-season grass species in temperate climate areas throughout the world, grown for forage with a high nutritive value. It is also widely used as a turf grass species with rapid establishment rate and excellent tolerance to traffic. These properties make the perennial ryegrass economically the most important species within the genus *Lolium* (Humphreys et al. 2006). Despite its many superior properties perennial ryegrass is sensitive to abiotic stresses. It exhibits poor winter survival under harsh and cold winters (Hulke et al. 2008), thus limiting its cultivation in certain areas. It was shown that perennial ryegrass responds to cold-acclimation, which increases its freezing tolerance (Ebdon et al. 2002). Cold acclimation is a multigenic quantitative trait associated with many physiological and biochemical changes within the plant cell (Hannah et al. 2005). Increase of water soluble sugar and proline concentration, as well as alteration of lipid composition and significant differentiation in profiles of protein accumulation is observed during cold acclimation in perennial ryegrass (Hoffman et al. 2010; Bocian et al. 2011). Proline acts as a free radical scavenger (Kaul et al. 2008), as osmoprotectant (Yoshiba et al. 1997), or as a protein-compatible hydrotrope (Srinivas and Balasubramanian 1995). The assumption was made that proline accumulation in plants under stress conditions has a protective function; however, correlation between proline accumulation and abiotic stress tolerance in plants is not always apparent (Szabados and Savouré 2010). Some data indicate that overproduction of proline in plants stabilises the cell membrane thereby preventing electrolyte leakage (Hayat et al. 2012). Moreover, low temperature can induce cell membrane lipid peroxidation (Campos et al. 2003), thus proline, as a free radical scavenger, could have an effect on membrane stability

G. Brazauskas (✉) · A. Aleliūnas · G. Statkevičiūtė · K. Jonavičienė · V. Kemešytė
Institute of Agriculture, Lithuanian Research Centre for Agriculture and Forestry,
Instituto a. 1, Akademija, 8344 Kėdainiai distr, Lithuania
e-mail: gintaras@lzi.lt

H. Budak, G. Spangenberg (eds.), *Molecular Breeding of Forage and Turf*,
DOI 10.1007/978-3-319-08714-6_2

maintenance. However, a lack of close association between freezing tolerance and proline content in perennial ryegrass and other species has been reported (Hoffman et al. 2010; Thomas and James 1993; Dionne et al. 2001). Cell membrane deterioration is the primary freeze-induced damage in plants, caused by cell dehydration, as it occurs when extracellular ice forms during freezing (Pearce 2001). Water potential of ice depends on the temperature, hence, the lower the temperature is the greater the damage is caused by the dehydration. Extracellular ice formation can also obstruct gas and solute exchange, while growing crystals can cause cell lysis (John et al. 2009). Freezing tolerant organisms have developed diverse mechanisms to survive subfreezing temperatures (Zachariassen and Kristiansen 2000) by minimising the damage caused by ice formation. Antifreeze proteins (AFPs) are considered to be one of the mechanisms to reduce freezing induced damage in plants and other kingdoms. Ice recrystallization activity (IRI) of AFPs is more common in freeze tolerant organisms such as plants (Middleton et al. 2012). As the common freeze-induced damage is cell membrane physical deteriorations, one of the most popular methods to measure the extent of the membrane injury is electrolyte leakage assay. This technique has been applied to quantify damages to cell membranes in various abiotic stress conditions (Bajji et al. 2001). The objective of this study was to evaluate perennial ryegrass association mapping population for variation in freezing tolerance related traits. This included estimation of winter survival of 210 perennial ryegrass accessions in the field, and further investigation of plant survival, electrolyte leakage and proline content in a subset of 76 genotypes under controlled-environment conditions.

2.2 Methods

2.2.1 Plant Material and DNA Extraction

Perennial ryegrass mapping panel consisted of 94 cultivars and 116 ecotypes. The ecotype group comprised 37 accessions from Lithuania, 3 accessions from Latvia, 2 from Poland, 9 accessions from Kaliningrad region (Russian Federation), 3 accessions from Slovakia and 62 accessions from Ukraine (Statkevičiūtė et al. 2014). All ecotype accessions were collected in their natural habitats. Cultivar group consisted of 89 cultivars developed in Europe, 4 in the USA and 1 in Japan. Plants were established from seeds and propagated in the greenhouse. Ploidy level of accessions was determined by Partec PA (Partec Gmbh, Germany) flow cytometer, and tetraploid plants were discarded from further analyses. Approximately 100 mg of pathogen-free leaf material of each genotype was harvested for genomic DNA extraction. DNA was isolated using a cetyltrimethyl ammonium bromide (CTAB) method according to the modified protocol of Lassner et al. (1989), supplemented with an additional chloroform wash step and RNA digestion with ribonuclease A. DNA concentration was quantified using a NanoDrop® 2000 (Thermo Fisher Scientific, USA) spectrophotometer.

2.2.2 Field Winter Survival Analysis

Ramet-based field collection of 210 perennial ryegrass accessions was established in the experimental fields of the Institute of Agriculture, Lithuanian Research Centre for Agriculture and Forestry (Akademija, Lithuania) in 2013. Single seed per accession was used for plant establishment in the greenhouse. Four ramets of each genotype were planted at 50 × 50 cm distances, using a randomised complete block design. Winter survival was evaluated in spring 2014 by visual scoring using a scale ranging from 1 (dead) to 9 (no visible damage) as described by Tyler et al. (1987).

2.2.3 Freezing Tolerance Testing Under Controlled Environment

Freezing tolerance test under controlled environment conditions was performed on a subset of 76 perennial ryegrass genotypes. Twenty days prior cold-acclimation plants were moved into phytotron, maintaining +20 °C, 450 μmol m^{-2} s^{-1} photosynthetically active radiation (PAR) and 80 % relative air humidity. Plants were cold-acclimated for 15 days prior the freezing test under controlled environment conditions at +5 °C, maintaining 200 μmol m^{-2} s^{-1} PAR, 12/12 h photoperiod and 80 % relative air humidity. After cold-acclimation period plants were divided into individual tillers, while leaves and roots were trimmed to 2 cm. Tillers were placed into test tubes of 10 ml, containing 1 ml of deionised water and subjected to freezing. Freezing tests were conducted in a walk-in freezing chamber PE 2412UY-LX (Angelantoni Industrie S.p.A, Italy) in the dark. Temperature was lowered from +5 °C to the desired treatment temperature at a rate of 1.2 °C h^{-1}. As temperature reached − 8 °C and − 12 °C test tubes of our examined genotypes were removed from the chamber and thawed on ice overnight at +4 °C. Later plants were replanted in cell trays and placed back to phytotron, while maintaining +20 °C, 450 μmol m^{-2} s^{-1} PAR and 80 % relative air humidity. Plant survival was scored after 3-week recovery period. Regrown plants were scored as survivors.

2.2.4 Electrolyte Leakage Assay

Membrane integrity of the plant cells was tested by measuring electrolyte leakage (EL) at − 8 °C and − 12 °C temperatures. Cold-acclimated plants were divided into ramets of single tillers, while leaves and roots were trimmed. Tillers were washed with deionised water, placed into glass test tubes containing 1 ml of deionised water and agitated for 15 min. Then they were washed again and wiped with paper towel. Four tillers per accession were placed into individual test tubes, containing 1 ml of deionised water and covered with parafilm. In total, three replicates per accession were prepared for each target temperature and randomly arranged in the test tube rack. Sample preparation work was conducted at +5 °C temperature. Prepared samples were subjected to freezing. When the target temperature of − 8 °C and − 12 °C was reached, racks were removed from freezing chamber and thawed on ice at +4 °C and 20 ml of deionised water was added into each tube. The tubes were shaken overnight at 120 rpm and then initial conductivity (C_{ini}) was measured by YSI 3100 (YSI Incorporated, USA) conductivity meter with YSI 3253 Glass Dip Cell. The total conductivity (C_{tot}) was measured after autoclaving the samples at 120 °C for 15 min. Electrolyte leakage (EL) was computed as $EL=(C_{ini})/(C_{tot}) \times 100$.

2.2.5 Proline Content Evaluation

Free proline content was determined in cold-acclimated plants prior to freezing application. Upper leaves were collected for each accession and immediately transferred to +70 °C preheated incubator and dried until the final weight of dry sample became stable. Dried leaves were ground with Retsch MM 400 (Retsch GmbH, Germany) mixer mill. Proline content was determined by the ninhydrin-based colorimetric assay (Ábrahám et al. 2010) assessing three replicates per accession.

2.2.6 Molecular Profiling by AFLP

AFLP profiling was used for the assessment of genetic diversity among the perennial ryegrass accessions. Genomic DNA (500 ng) was digested with Fast Digest® *Taq*I and *Ase*I restriction endonucleases (Thermo Scientific, Lithuania). *Taq*I adapter was prepared by combining 25 µg oligo_T_top and 22 µg oligo_T_bottom (Table 2.1) and deionised water was added to total volume of 100 µl. Prepared solution was incubated at 95 °C for 5 min and then slowly cooled for 2 h. *Ase*I adapter was prepared in the same manner, except 2.6 µg of oligo_A_top and 2 µg of oligo_A_bottom were used. The adapters were added to the digested DNA mixture (as described above) to final concentrations of 50 pM (*TaqI* adapter) and 5 pM

Table 2.1 List of the oligos for AFLP analyses

Oligo ID	Oligo sequence
oligo_A_bottom	5'- TAGGTACGCAGTC
oligo_A_top	5'- CTCGTAGACTGCGTACC
oligo_T_top	5'- GACGATGAGTCCTGAC
oligo_T_bottom	5'- CGGTCAGGACTCAT
oligo_Pre_amp_F	5'- CTCGTAGACTGCGTACCTAAT
oligo_Pre_amp_R	5'- GACGATGAGTCCTGACCGA
IR800Ase + ACG	5'- GACTGCGTACCTAATACG
IR800Ase + ACC	5'- GACTGCGTACCTAATACC
Taq + CAC	5'- GATGAGTCCTGACCGACAC
Taq + CTA	5'- GATGAGTCCTGACCGACTA
Taq + CAA	5'- GATGAGTCCTGACCGACAA

(*Ase*I adapter) and ligated to restricted genomic DNA fragments. Preamplification was carried out using 10 μl tenfold diluted ligation mixture, 0.5 μM oligo_Pre_amp_F and 0.5 μM oligo_Pre_amp_R preamplification primers and 1x DreamTaq PCR Master Mix (Thermo Scientific, Lithuania) in 50 μl total volume using following PCR program with 20 cycles of each: 94 °C for 30 s, 55 °C for 30 s, 72 °C for 1 min. Selective amplification was performed on 5 μl 20-fold diluted preamplification PCR product, as PCR mixture consisted of 1x DreamTaq PCR buffer, 0.25 mM dNTP, 1.25 units of DreamTaq DNA polymerase, 0.05 μM IRDye®800-labelled forward primer and 0.3 μM reverse primer. Total volume of the mixture was adjusted to 10 μl by adding deionised water. Three primer combinations, harbouring three selective bases were used for selective amplification in AFLP: IR800AseACG + TaqCAC, IR800AseACC + TaqCTA and IR800AseACC + TaqCAA (Table 2.1). PCR program for selective amplification included 12 cycles of each: 94 °C for 30 s, initial annealing temperature of 65 °C (with decrement of −0.7 °C each cycle) for 30 s, 72 °C for 1 min, followed by 24 cycles with 94 °C for 30 s, 56 °C for 30 s, 72 °C for 1 min. PCR products were separated on 6.5 % KBPLUS Gel Matrix (LI-COR, Inc., USA) and detected by the DNA Analyzer 4300 (LI-COR, Inc., USA). Fragments were scored using SAGA Generation 2 software (LI-COR, Inc., USA). AFLP analysis was conducted in duplicate to check for the consistency between replicates.

2.2.7 Population Structure and Relative Kinship

An admixture model in STRUCTURE software package (Pritchard et al. 2000) was used to infer population structure. Ten independent simulations were run by setting the number of populations (k) from 1 to 10, with 100,000 burn-in time and 100,000 iterations of Markov chain convergence for each run. The subpopulation number was determined using methodology developed by Evanno et al. (2005). Pairwise kinship matrix was calculated by SPAGeDi1-4c (Hardy and Vekemans 2002) using AFLP dataset. All negative kinship values between individuals were set to zero.

2.2.8 Statistic Analysis

Analysis of variance (ANOVA) with *post hoc* Tukey HSD test and Spearman's rank correlation coefficients were calculated using STATISTICA 7 (StatSoft, Inc.).

2.3 Results

Winter survival under field conditions was evaluated for 210 perennial ryegrass genotypes. Analysis of variance revealed that one replicate block in our field collection differed significantly ($p < 0.05$) from the other three in winter survival, so it was excluded from further analyses. ANOVA *post hoc* Tukey HSD test on winter survival revealed significant ($p<0.0001$) differences between ecotype and cultivar groups. Plants within ecotype group showed better winter survival performance compared to cultivar group. The mean winter survival score was 4.71 and 2.76 for ecotype and cultivar groups, respectively. In general, perennial ryegrass accessions demonstrated good winter survival with only 41 plant out of 630 being lost due to poor overwintering. A subset of 76 perennial ryegrass genotypes was further examined for freezing tolerance in controlled-environment experiment. Freezing tolerance was defined as percentage of tiller survival (PTS) in controlled freezing test. PTS ranged from 0 to 100 % (mean 55.76 %±3.67 %) and from 0 to 91.67 % (mean 9.99 %±2.01 %) at −8 °C and −12 °C, respectively. Twenty five freeze-susceptible (PTS at −8 °C<44 %) and 5 freeze-tolerant (PTS at −12 °C>44 %) genotypes were identified in this subset. While significant differences in freezing tolerance were observed between individual genotypes, ANOVA *post hoc* Tukey HSD test revealed no differences in PTS between ecotype and cultivar groups at both tested temperatures. Investigated perennial ryegrass accessions significantly ($p<0.001$) differed in proline content (PC), as well as in electrolyte leakage (EL) (Table 2.2). PC ranged from 0.47 to 6.15 mg g^{-1} DW, while EL varied from 6.09 to 25.80 % (mean

Table 2.2 Electrolyte leakage (EL) at −8 °C and −12 °C, percentage tiller survival (PTS) at −8 °C and −12 °C, proline content (PC) and winter survival (FWS) variation among perennial ryegrass accessions

Trait	Ecotypes				Cultivars			
	Mean	Min	Max	SD[a]	Mean	Min	Max	SD[a]
EL −8 °C	14.27	6.09	25.80	5.23	15.83	7.82	25.06	4.82
EL −12°C	38.91	22.27	57.06	9.14	43.09	23.85	63.51	11.48
PTS −8°C	53.46	0.00	100.00	5.32	58.75	0.00	100.00	4.80
PTS −12°C	10.30	0.00	91.67	3.15	9.60	0.00	53.33	2.62
PC	2.55	0.47	6.15	1.36	2.52	0.51	5.84	1.34
FWS	4.71	0.00	9.00	2.77	2.76	0.00	9.00	2.36

[a] Standard deviation

Table 2.3 Phenotypic correlation between electrolyte leakage (EL) at −8 °C and −12 °C, percentage tiller survival (PTS) at −8 °C and −12 °C, proline content (PC) and field winter survival (FWS)

Trait	EL −8°C	EL −12°C	PTS −8°C	PTS −12°C	PC
	rho[a]	rho	rho	rho	rho
EL −12°C	0.43***	–	–	–	–
PTS −8°C	−0.40***	−0.21 ns	–	–	–
PTS −12°C	−0.14 ns[b]	−0.49***	0.46***	–	–
PC	−0.06 ns	0.02 ns	−0.20 ns	0.03 ns	–
FWS	−0.21 ns	−0.48***	0.01 ns	0.03 ns	−0.06 ns

*** Significant ($p<0.001$)

[a] Spearman's rank correlation coefficient

[b] Non significant ($p>0.05$)

15.17 %±0.58 %) at −8 °C and from 22.27 to 63.51 % (mean 41.10 %±1.23 %) at −12 °C between individual accessions. No significant ($p>0.05$) differences were detected in PC between ecotype and cultivar groups. ANOVA *post hoc* Tukey HSD test also revealed no significant ($p>0.05$) differences in EL at −8 °C between ecotype and cultivar groups, however difference in EL at −12 °C between the groups was significant ($p<0.05$), averaging 43.94 %±1.94 % and 39.09 %±1.42 % for cultivar and ecotype groups, respectively. No significant ($p>0.05$) differences in EL between plants in ecotype group regarding to the place of origin at both tested freezing temperatures was detected. Spearman's rank correlation coefficients between investigated individual traits were calculated (Table 2.3). Significant moderate negative correlations were observed between EL at −12 °C and PTS at −12 °C (rho=−0.49, $p<0.001$), as well as EL at −8 °C negatively correlated with PTS at −8 °C (rho=−0.40, $p<0.001$). Moderate positive correlation (rho=0.43, $p<0.001$) was found between EL at −8 °C and EL at −12 °C, as well as PTS at −8 °C and PTS at −12 °C (rho=0.46, $p<0.001$). There was no correlation between PC and any other examined trait. No significant ($p>0.05$) correlation was observed between winter survival and freezing-tolerance under controlled environment conditions. Population structure and kinship was assessed with AFLP markers. AFLP markers are known to show random distribution across genomes which makes them markers of choice for population structure examination. AFLP analysis scored a total of 227 bands. Majority of the bands were polymorphic (98.67 %). Inbreeding coefficient was low with a value of 0.068. Methodology of Evano et al. (2005) was used to define the number of subpopulations. The Q value matrix with k=2 was generated, which indicates two clusters in our perennial ryegrass panel. The STRUCTURE analysis revealed that most ecotype accessions from Ukraine fall into the subpopulation 1 while ecotypes from Lithuania, Latvia and Kaliningrad region correspond to subpopulation 2. AFLP data showed remarkable heterogeneity among cultivars with accessions nearly-equally distributed between both subpopulations. No obvious kinship was found in perennial ryegrass panel. Pairwise kinship estimates were low with 49.49 % of the values equal to zero, while 82.71 % were lower than 0.05.

2.4 Discussion

Winter survival is a complex trait. Although freezing-tolerance of the plants is considered as the major component of winter survival (Teutonico et al. 1995), the absence of significant correlation between PTS under controlled-environment conditions and FWS under field conditions in this experiment indicates the existence of other major factors contributing to winter survival performance of the perennial ryegrass. Furthermore, the absence of correlation between the traits could reflect the differences in cold-acclimation, as relatively long and warm autumn in 2013 could have had an impact on the physiological condition of the plants in the field. In general, ecotypes demonstrated better FWS performance than cultivars. It indicates that the genetic potential for freezing tolerance has not been fully exploited in perennial ryegrass cultivars. A subset of 76 genotypes was screened for direct and indirect freezing tolerance parameters in controlled-environment experiment. EL assay was used to assess cell membrane damage and very low variation was observed between replicates using this method (data not shown). No differentiation in freezing tolerance parameters between ecotype and cultivar groups was detected, except for EL at − 12 °C. EL is considered as indirect freezing tolerance parameter (Espevig et al. 2011). Ebdon et al. (2002) revealed that LT_{50} derived from EL curves correlate with winter survival estimates for both hardened and non-hardened perennial ryegrass plants, nonetheless we found only moderate negative correlations between EL and PTS at both tested temperatures. Most studies use LT_{50} or EL_{50} values to define plant freezing tolerance and EL. Estimation of these values are labour-intensive and time-consuming, thus problematic for numerous sample sizes. A freezing test at fixed-point temperature was chosen to screen a larger number of samples in the current population. Cell membrane stability is maintained by the accumulation of solutes, such as free proline, during cold acclimation (Zhang et al. 2010). Substantial variation for free proline content was observed in our perennial ryegrass panel; however, no correlation could be detected with other freezing tolerance parameters. While some studies indicate, that proline plays a role in cell protection against detrimental effects during freezing (Xin and Browse 2000), it's role is still controversial. Although significantly higher proline content in freezing tolerant accessions of perennial ryegrass as compared to freezing-susceptible genotypes was detected after 14 days of cold-acclimation, the differences were not apparent a week later (Hoffman et al. 2010). Thomas and James (Thomas and James 1993) compared freezing tolerance of compost- and sand-grown perennial ryegrass. They found the latter to be as hardy as grown in compost, despite having a tenth fraction of the proline content. Those findings altogether make evidence that there is no strong association between proline content and freezing tolerance.

AFLP markers revealed two subpopulations in the population of 210 genotypes and low kinship between individuals. High variation for freezing tolerance and minimum kinship between genotypes indicate the suitability of the population for association mapping of freezing tolerance related traits, while existence of two subpopulations implements the use of population structure controlled model for marker-trait association analysis.

Acknowledgments This study was funded by the Research Council of Lithuania, grant No. MIP-032/2012 (FUMAG).

References

Ábrahám E, Hourton-Cabassa C, Erdei L, Szabados L (2010) Methods for determination of proline in plants. In: Sunkar R (ed) Plant stress tolerance. Humana Press, New York, pp 322–324

Bajji M, Kinet JM, Lutts S (2001) The use of the electrolyte leakage method for assessing cell membrane stability as a water stress tolerance test in durum wheat. Plant Growth Regul 00:1–10

Bocian A, Kosmala A, Rapacz M, Jurczyk B, Marczak Ł, Zwierzykowski Z. (2011) Differences in leaf proteome response to cold acclimation between *Lolium perenne* plants with distinct levels of frost tolerance. J Plant Physiol 168:1271–1279

Campos PS, Quartin V, Ramalho JC, Nunes MA (2003) Electrolyte leakage and lipid degradation account for cold sensitivity in leaves of *Coffea sp.* plants. J Plant Physiol 160:283–292

Dionne J, Castonguay Y, Nadeau P, Desjardins Y (2001) Amino acid and protein changes during cold acclimation of green-type annual bluegrass (*Poa annua* L.) ecotypes. Crop Sci 41:1862–1870

Ebdon JS, Gagne RA, Manley RC (2002) Comparative cold tolerance in diverse turf quality genotypes of perennial ryegrass. Hort Sci 37:826–830

Espevig T, Dacosta M, Hoffman L, Aamlid TS, Tronsmo AM, Clarke BB, Huang B (2011) Freezing tolerance and carbohydrate changes of two agrostis species during cold acclimation. Crop Sci 51:1188–1197

Evanno G, Regnaut S, Goudet J (2005) Detecting the number of clusters of individuals using the software structure: a simulation study. Mol Ecol 14:2611–2620

Hannah MA, Heyer AG, Hincha DK (2005) A global survey of gene regulation during cold acclimation in *Arabidopsis thaliana*. PLoS Genet 1:e26

Hardy OJ, Vekemans X (2002) SPAGeDi: a versatile computer program to analyse spatial genetic structure at the individual or population levels. Mol Ecol Notes 2:618–620

Hayat S, Hayat Q, Alyemeni MN, Wani AS, Pichtel J, Ahmad A (2012) Role of proline under changing environments: a review. Plant Signal Behav 7:1456–1466

Hoffman L, DaCosta M, Ebdon JS, Watkins E (2010) Physiological changes during cold acclimation of perennial ryegrass accessions differing in freeze tolerance. Crop Sci 50:1037–1047

Hulke BS, Watkins E, Wyse DL, Ehlke NJ (2008) Freezing tolerance of selected perennial ryegrass (*Lolium perenne* L.) accessions and its association with field winterhardiness and turf traits. Euphytica 163:131–141

Humphreys MW, Yadav RS, Cairns AJ, Turner LB, Humphreys J, Skøt L (2006) A changing climate for grassland research. New Phytol 169:9–26

John UP, Polotnianka RM, Sivakumaran KA, Chew O, Mackin L, Kuiper MJ, Talbot JP, Nugent GD, Mautord J, Schrauf GE, Spangenberg GC (2009) Ice recrystallization inhibition proteins (IRIPs) and freeze tolerance in the cryophilic Antarctic hair grass *Deschampsia antarctica* E. Desv. Plant Cell Environ 32:336–348

Kaul S, Sharma SS, Mehta IK (2008) Free radical scavenging potential of L-proline: evidence from in vitro assays. Amino Acids 34:315–320

Lassner MW, Peterson P, Yoder JI (1989) Simultaneous amplification of multiple DNA fragments by polymerase chain reaction in the analysis of transgenic plants and their progeny. Plant Mol Biol Rep 7:116–128

Middleton AJ, Marshall CB, Faucher F, Bar-Dolev M, Braslavsky I, Campbell RL, Walker VK, Davies PL (2012) Antifreeze protein from freeze-tolerant grass has a beta-roll fold with an irregularly structured ice-binding site. J Mol Biol 416:713–724

Pearce SR (2001) Plant Freezing and Damage. Ann Bot 87:417–424

Pritchard JK, Stephens M, Donnelly P (2000) Inference of population structure using multilocus genotype data. Genetics 155:945–959

Srinivas V, Balasubramanian D (1995) Proline is a protein-compatible hydrotrope. Langmuir 11:2830–2833

Statkevičiūtė G, Aleliūnas A, Kemešytė V, Pašakinskienė I, Brazauskas G (2014) AFLP analysis of genetic diversity in an association mapping panel of *Lolium perenne* L. In: Sokolović D, Huyghe C, Radović J (eds) Quantitative traits breeding for multifunctional grasslands and turf. Springer, Netherlands

Szabados L, Savouré A (2010) Proline: a multifunctional amino acid. Trends Plant Sci 15:89–97

Teutonico RA, Yandell B, Satagopan JM, Ferreira ME, Palta JP, Osborn TC (1995) Genetic analysis and mapping of genes controlling freezing tolerance in oilseedBrassica. Mol Breed 1:329–339

Thomas H, James AR (1993) Freezing tolerance and solute changes in contrasting genotypes of *Lolium perenne* L. acclimated to cold and drought. Ann Bot 72:249–254

Tyler BF, Hayes JD, Davies WE (eds) (1987) Collection, characterization and utilization of genetic resources of temperate forage grass and clover. IBPGR, Rome, pp 1–65

Xin Z, Browse J (2000) Cold comfort farm: the acclimation of plants to freezing temperatures. Plant Cell Environ 23:893–902
Yoshiba Y, Kiyosue T, Nakashima K, Yamaguchi-Shinozaki K, Shinozaki K (1997) Regulation of levels of proline as an osmolyte in plants under water stress. Plant Cell Physiol 38:1095–1102
Zachariassen KE, Kristiansen E (2000) Ice nucleation and antinucleation in nature. Cryobiology 41:257–279
Zhang C, Fei SZ, Arora R, Hannapel DJ (2010) Ice recrystallization inhibition proteins of perennial ryegrass enhance freezing tolerance. Planta 232:155–164

Chapter 3
Reciprocal Full-sib Recurrent Selection and Tools for Accelerating Genetic Gain in Apomictic *Brachiaria*

M. L. Worthington and J. W. Miles

ASGR	Apospory-specific genomic region
CIAT	International Center for Tropical Agriculture
EMBRAPA	Brazilian Enterprise for Agricultural Research
MAS	Marker assisted selection
QTL	Quantitative trait loci
RAPD	Random amplified polymorphic DNA
RFRS	Reciprocal full-sib recurrent selection
RRS	Reciprocal recurrent selection
RS-SCA	Recurrent selection on specific combining ability
SCAR	Sequence characterized amplified region
SNP	Single nucleotide polymorphism

3.1 Introduction

Four economically important forage grass species in the genus *Brachiaria* are native to East and Central Africa. After their introduction to tropical America in the 1960s, these species quickly gained popularity due to their rusticity and tolerance to poor soils and aluminum toxicity (Miles et al. 2004). Brachiariagrass cultivars are planted on 99 Mha in Brazil alone, accounting for 85 % of sown pasture land (Jank et al. 2014). *Brachiaria* is also planted extensively in other Latin American countries and is grown by small-holder farmers in Africa and Asia (Stür et al. 2006; Mutimura and Everson 2011; Khan et al. 2014).

Brachiaria brizantha (A. Rich.) Stapf (palisadegrass), *B. decumbens* Stapf (signalgrass), and *B. humidicola* (Rendle) Schweick (koroniviagrass), the most economically important grasses of the genus *Brachiaria*, are polyploid apomicts.

M. L. Worthington (✉) · J. W. Miles
International Center for Tropical Agriculture (CIAT), A.A. 6713, Cali, Colombia
e-mail: M.Worthington@gmail.com

H. Budak, G. Spangenberg (eds.), *Molecular Breeding of Forage and Turf*,
DOI 10.1007/978-3-319-08714-6_3

A fourth commercial species of lesser importance is the diploid sexual *B. ruziziensis*. The term "apomixis" can describe any form of asexual reproduction, but has come to be used to refer to asexual reproduction through seed (Asker and Jerling 1992). There are several different recognized forms of apomixis (Asker and Jerling 1992), but apospory is the most common type found in *Brachiaria* and other warm-season forage and turf-grasses including guineagrass (*Panicum maximum* Jacq.), buffelgrass [*Pennisetum ciliare* (L.) Link. (syn. *Cenchrus ciliaris* L.)], and bahiagrass (*Paspalum notatum* Flüggé) (Miles 2007).

Aposporous female gametophytes are formed from somatic nucellar cells and develop without fertilization into unreduced embryos genetically identical to the mother plant, thereby faithfully reproducing (cloning) the maternal genotype in seedling progenies. Because an unreduced apomictic embryo sac develops from a somatic nucellar cell following only two mitoses, it has only four nuclei (one ovule, two synergids, and a single polar nucleus). Mature apomictic embryo sacs can, thus, be distinguished from normal eight-nucleate sexual sacs using pistil clearing techniques that facilitate embryo sac analysis (Young et al. 1979).

In *Brachiaria* and other warm season grasses, apospory is controlled by a single dominant Mendelian factor, denoted the "apospory-specific genomic region" (ASGR) (Ozias-Akins and Dik 2007). Diploid accessions are exclusively sexual, while the vast majority of polyploid accessions are apomicts. Tetraploid apomicts possess a simplex genotype (Aaaa) at the ASGR, while sexual diploids (aa), and tetraploids (aaaa) are homozygous recessive (Savidan 1983). However, not all plants with the simplex genotype are obligate apomicts; sexual and apomictic embryo sacs are often found on the same plant or even in the same ovule. While facultative apospory is common, the genetic control of this trait is not understood.

Apomixis allows breeders to identify and faithfully propagate by seed superior heterozygous genotypes without the disadvantages of vegetative propagation or the expense and complexity of hybrid seed production from inbred parental lines. However, despite the many potential benefits of apomictic reproduction, the difficulty of achieving genetic recombination in polyploid *Brachiaria* germplasm originally posed a strong barrier to breeding. Although many breeders and researchers still perceive apomixis as an impediment to plant breeding (Vigna et al. 2011; Silva et al. 2013), barriers to recombination have been overcome in two *Brachiaria* gene pools (*B. humidicola* and the interfertile *B. brizantha–B. decumbens–B. ruziziensis* agamic complex). This chapter provides a brief summary of current breeding methods in the *B. brizantha–B. decumbens–B. ruziziensis* agamic complex (hereafter referred to as *Brachiaria*) and outlines a proposal for a reciprocal full-sib recurrent selection (RFRS) procedure suitable for use in apomictic species.

3.2 Recurrent Selection for Specific Combining Ability

Brachiaria breeding was first enabled by the development of a tetraploid sexual biotype of *Brachiaria ruziziensis* by colchicine treatment of a natural sexual diploid accession in the 1980s (Swenne et al. 1981). Microsporogenesis occurs normally in apomictic plants. Therefore, a tetraploidized *B. ruziziensis* genotype was used as the female parent and exposed to pollen of apomictic genotypes of *B. brizantha* and *B. decumbens* at CIAT and Empresa Brasileira de Pesquisa Agropecuária (EMBRAPA; Brazilian Enterprise for Agricultural Research) in the late 1980s to form the basis of today's breeding programs. Breeders were able to identify sexual and apomictic genotypes among the progeny of these initial crosses using embryo sac analysis (Young et al. 1979; Valle and Savidan 1996) and progeny tests. Embryo sac analysis and progeny tests can be used to reliably screen relatively large populations, but both processes are expensive and time consuming. The reproductive mode of a given genotype cannot be assessed until flowering using embryo sac analysis, while progeny tests cannot be used to determine the reproductive mode of a given genotype until its offspring can be evaluated in field or pot trials. Thus, in the absence of a screening tool or molecular marker that could be efficiently and reliably used to evaluate seedlings for reproductive mode, CIAT and EMBRAPA developed genetically diverse and fully sexual breeding populations by recombining sexual hybrids from early crosses for recurrent selection.

Because *Brachiaria* cultivars are true breeding heterozygous clones, an ideal breeding scheme should deliberately exploit heterosis by accumulating heterotic effects over cycles of recurrent selection. Miles (Miles 2007) proposed recurrent selection for specific combining ability (RS-SCA) as an appropriate method for increasing heterotic effects in a sexual *Brachiaria* breeding population in the absence of a reliable marker that could be used to screen seedlings for reproductive mode. Originally proposed in maize (Hull 1945), RS-SCA is a three-step process where (1) plants from a recurrent selection population are crossed to a single "tester" genotype, (2) superior parents are chosen on the basis of the performance of their testcross progeny, and (3) the selected parental genotypes are recombined to form the basis of the next cycle of selection.

In year 1 of the CIAT *Brachiaria* RS-SCA breeding scheme, sexual clones are crossed as females with an elite non-inbred apomictic pollen parent to generate progenies segregating 1:1 for reproductive mode. Hybrid (testcross) progenies are evaluated in field trials as vegetatively replicated spaced plants in year 2. Sexual maternal plants are maintained and selected as parents for the next cycle of RS-SCA on the basis of their testcross progeny performance. In the final year of the cycle, the selected sexual clones are recombined in an isolated, open-pollinated crossing block (Fig. 3.1). Meanwhile superior apomictic individuals selected from the testcross progenies are further evaluated as candidate cultivars.

B. decumbens cultivar "Basilisk" was chosen as the apomictic tester for RS-SCA in the CIAT program, based on its general combining ability with clones from the sexual recurrent selection process (Miles 2007). Basilisk, a selection from a natural germplasm collection, is a highly successful commercial cultivar with excellent tolerance to aluminum toxicity and poor soils, but it is very susceptible to spittlebug damage (Jank et al. 2014). Thus, Basilisk was also seen as a logical complement to the sexual recurrent selection population, which had been subjected to several cycles of intense selection for spittlebug resistance but had relatively poor edaphic adaptation (Miles 2007; Miles et al. 2006).

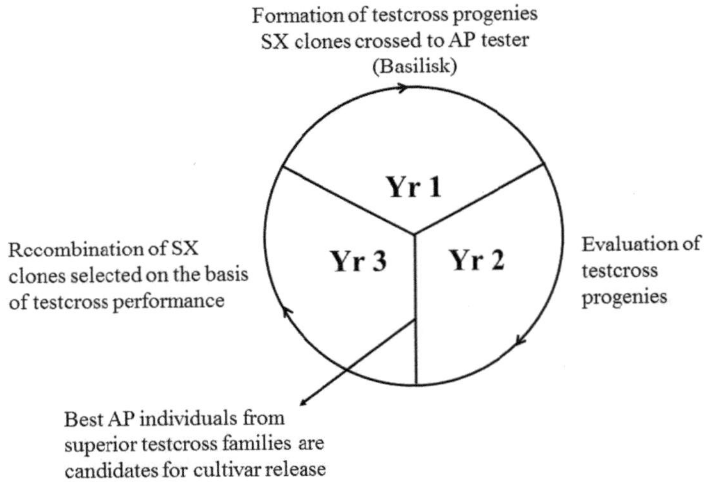

Fig 3.1 Breeding scheme for recurrent selection on specific combining ability (RS-SCA) in a synthetic sexual tetraploid *Brachiaria* breeding population. *SX* sexual, *AP* apomictic (Miles 2007)

The first RS-SCA testcross progenies were generated in 2005. Approximately 500 sexual clones from the synthetic sexual population were transplanted on 5-m centers in an established Basilisk pasture. The self-incompatibility of sexual *Brachiaria* genotypes (Ngendahayo et al. 1988) and wide-spacing between sexual plants assured that most open-pollinated seeds would be hybrids with the Basilisk tester. The following year, 2250 genotypes from 233 testcross families were evaluated as single vegetative replicates at two field sites. Thirty-one parents were selected for recombination and 168 hybrid genotypes were chosen for further evaluation. As expected, approximately 50 % of selected hybrid genotypes were discarded after progeny tests revealed that they were sexually reproducing.

A random amplified polymorphic DNA (RAPD) marker linked to the ASGR was first identified in 1995 (Pedraza Garcia 1995). This marker, named "N14," was converted to a sequence characterized amplified region (SCAR) marker and first used to screen hybrid testcross progeny seedlings in 2009 during the second cycle of RS-SCA. Approximately 3000 putative sexual testcross seedlings were discarded on the basis of marker scores before transplanting to the field, ensuring that the majority of the 3000 remaining hybrid genotypes screened in the field were apomicts. Concerns about the accuracy of N14 stalled its use in the CIAT breeding program for 14 years, but progeny tests in 2010 confirmed that the marker predicted apomictic reproduction with over 90 % accuracy (Miles, unpublished data). The use of N14 to screen hybrid seedlings has allowed breeders to evaluate a greater number of apomictic genotypes in initial field trials each cycle and provided considerable savings in time and money used to establish sexual hybrids unsuitable for further use in the breeding program. N14 has now been used to screen over 12,000 hybrid seedlings for reproductive mode during cycles 2 and 3 of RS-SCA, and will be used to screen the fourth cohort of testcross genotypes generated by RS-SCA in 2015.

3.3 Reciprocal Recurrent Selection

Comstock (1949) proposed reciprocal recurrent selection (RRS) as a method simultaneously to improve general and specific combining ability in two complementary populations that exhibit heterosis in their crosses. The three-generation breeding cycle proposed in maize involves the (1) formation of hybrid progenies by crossing selected plants from Population A to a random set of plants from Population B and vice versa, (2) evaluation of half-sib families and checks at multiple field sites, and (3) recombination of remnant selfed seed from the progenitors of superior families of Population A and B in separate nurseries. This method has been shown to be nearly as effective as RS-SCA for loci at which there is overdominance and superior for loci at which there is partial dominance (Comstock et al. 1949). Therefore, RRS should result in accelerated gain compared to RS-SCA in the CIAT *Brachiaria* breeding program.

Several modifications to RRS have been suggested, including reciprocal full-sib recurrent selection (RFRS), which involves the formation and evaluation of full-sib families through paired crosses between plants in Populations A and B (Hallauer and Eberhart 1970). Another modified form of RRS has been proposed for Timothy (*Phleum pratense* L.), an outcrossing forage species bred using a "clone and strain synthesis" approach that minimizes inbreeding depression (Tanaka et al. 2011). Forms of RRS have also been implemented in a number of clonally propagated tropical tree species, including cocoa (*Theobroma cacao* L.) (Pokou et al. 2009), coffee (*Coffea canephora* Pierre) (Leroy et al. 1993), oil palm (*Elaeis guineensis* Jacq) (Bakoumé et al. 2010), and eucalyptus (*Eucalyptus* L'Hér.) (Baudouin et al. 1997). Although clonally propagated annual crops would be ideal candidates for RRS (Miles 2007; Hull 1945; Grüneberg et al. 2009), RRS is not currently being implemented in any of these species to the best of our knowledge.

Miles and Valle (1996) first considered the possibility of improving populations segregating for apomixis almost 20 years ago, but ultimately decided that the expense and logistics of determining the reproductive mode of large numbers of genotypes was too big an obstacle. Miles (2007) revisited the idea a decade later, but concluded that there was still no simple and reliable screening tool available that could practically be used to distinguish apomictic from sexual genotypes in a segregating population. While N14 is not completely accurate in distinguishing between sexually reproducing plants and apomicts, this marker allows for sufficient management of populations segregating for apomixis to justify the initiation of RRS. Moreover, the increasing availability and decreasing cost of next generation sequencing tools should enable the discovery of single nucleotide polymorphism (SNP) markers even more closely linked to the ASGR in the near future. Although EMBRAPA scientists have referred to their *Brachiaria* breeding scheme as "reciprocal recurrent selection" (Jank et al. 2014; Barrios et al. 2013), the methods described by Barrios et al. (2013) are equivalent to the RS-SCA scheme published by Miles (2007). To the best of our knowledge, true RRS has never been implemented in *Brachiaria* or any apomictic species and this chapter provides the first feasible proposal for modified RRS suitable for use in apomictic species.

Reciprocal recurrent selection in *Brachiaria* differs from other previously described breeding schemes in the sense that at least one population, and ideally both populations, will have to be segregating for reproductive mode. Every genotype in the breeding pipeline, including all parents from both populations (A and B) and all reciprocal hybrid progeny, must be genotyped for reproductive mode using N14 or another molecular marker for apomixis at seedling stage. Sexual plants from Population A will be crossed to apomictic plants from Population B (and an equal number of sexual plants from B to apomicts from A) to establish two sets of reciprocal full-sib progenies. The sexual and apomictic parents of selected full-sib families from each population will be recombined in isolation from the complementary population. Seed harvested from sexual plants in these recombination blocks will give rise to new populations segregating for reproductive mode.

RFRS offers two major benefits over RRS conducted on half-sib progeny. The utilization of paired-crosses between apomictic and sexual genotypes at both the testcross generation and intrapopulation recombination stages would facilitate the simple maintenance of roughly equal numbers of apomictic and sexual plants over breeding cycles. The extensive pedigree information made available in RFRS would also facilitate the use of best linear unbiased predictions (BLUP) to estimate the genetic values of interpopulation hybrid genotypes and the breeding values of both female and male parental genotypes, which are not phenotyped during the evaluation of full-sib progeny in relevant field sites (Piepho et al. 2008; Viana et al. 2014). Thus, RFRS will enable the selection of both sexual and apomictic plants with superior breeding values in each of the two complementary populations.

A detailed proposed procedure for RFRS in two reciprocal, outcrossing breeding populations, Population A and Population B, each segregating for ASGR is outlined in detail in Fig. 3.2. The numbers of plants and progenies indicated were chosen based on the practical limitations of CIATs current breeding program, with some consideration for maintaining high selection intensity while minimizing genetic drift.

Year 1: Formation of full-sib families by paired crosses between sexual and apomictic plants from complementary populations A and B (i.e., $SX_A \times AP_B$ and $SX_B \times AP_A$).

Reproductive mode of approximately 200 seedlings from each population will be assessed with a reliable molecular marker. Each population should have approximately 100 sexual genotypes and 100 apomictic genotypes. Grow out two vegetative replicates of each genotype in pots to initiate flowering. Make paired crosses between sexual plants from Population A and apomictic plants from Population B ($SX_A \times AP_B$) and between sexual plants from Population B and apomictic plants from Population A ($SX_B \times AP_A$) at random, but taking into account more or less synchronous flowering between the two members of each pair. When possible, parental genotypes will be used in multiple crosses in order to obtain better estimates their general combining ability. If each genotype is represented in two crosses, approximately 400 full-sib families will be generated.

Year 2: Field evaluation of full-sib families.

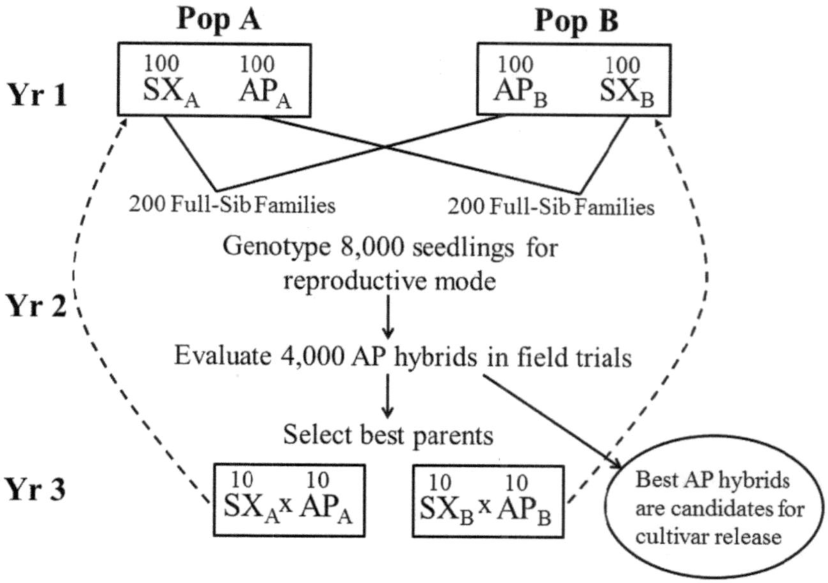

Fig 3.2 Proposed scheme for reciprocal full-sib recurrent selection (RFRS) (Hallauer and Eberhart 1970) in two complementary tetraploid *Brachiaria* breeding populations segregating for apomixis. *SX* sexual, *AP* apomictic

Genotype for reproductive mode approximately 8000 seedlings from the 400 full-sib families generated in year 1. Approximately 50 % of seedlings from each full-sib family (i.e., ~ 10 in this example) are expected to be sexually reproducing. Discard all sexual plants and establish at least two vegetative replicates of each apomictic genotype as spaced plants in field trials in representative test environments. Visually evaluate each plot and select superior (apomictic) hybrid individuals and parents with excellent specific and general combining ability using BLUP. Superior apomictic individuals become candidates for cultivar status and are subjected to advanced screening for various traits of interest.

Year 3: Recombine selected parental clones within populations A and B

Recombine, within the two reciprocal populations, the parental clones selected on the basis of BLUP in year 2. Parental clones have to be maintained, as living perennial plants, from the first year of the breeding cycle. Approximately, 10 sexual and 10 apomictic clones are selected from each population and pair-crossed to form approximately 20 full-sib families. The recombined seedlings produced from crosses between apomicts and sexual plants within Populations A and B form the basis of the following breeding cycle.

Reciprocal full-sib recurrent selection should fit nicely into the breeding pipeline at CIAT. As with RS-SCA, each breeding cycle takes 3 years to complete. Field evaluations of testcross progenies can be conducted following the same protocols used currently in RS-SCA. While RFRS requires an additional genotyping step, only 400 parental plants from the complementary populations need to be genotyped for reproductive mode in year 1 of each breeding cycle. Over 6000 testcross hybrids are already genotyped during year 2 of each breeding cycle, so the process of genotyping 400 parental plants in year 1 of the breeding cycle requires a relatively modest additional investment of time and resources. The most drastic change to current procedures proposed in this breeding scheme is the use of paired crosses. However, paired-crossing should be relatively simple to implement by placing synchronously flowering potted apomictic and sexual plant pairs together in pollen exclusion cages in the greenhouse. The high degree self-incompatibility in sexual *Brachiaria* plants (Ngendahayo et al. 1988) should ensure that the majority of seed harvested from sexual plants are products of a cross with the apomictic pollen donor. Any unintentional selfs are eliminated when testcross progenies are screened for reproductive mode.

3.4 Heterotic Groups

The first important step toward implementing RRS in *Brachiaria* is the establishment of complementary heterotic groups. As with any breeding program, the alleles present in the chosen parental material limit the future achievements. Parental material should both have the best performance possible for traits of interest to the breeding program and high levels of genetic diversity. For RRS, it is also important that the two complementary breeding populations display a high level of heterosis for characters of interest when crossed. Heterotic patterns in breeding populations can be developed over cycles of RRS (Labate et al. 1997). However, the exploitation of existing germplasm pools and natural heterotic patterns can accelerate gain and improve early cycle hybrid performance.

Heterotic patterns between germplasm pools can often be inferred on the basis of their genetic distance. Studies measuring hybrid performance in inter-pool and intra-pool crosses show that genetic distance is a good predictor of average hybrid performance between germplasm pools, but not a very strong predictor of the performance of hybrid progeny from specific crosses (Melchinger et al. 1992; Link et al. 1996; Cheres et al. 2000). Genetic distance between natural germplasm pools has been used to identify heterotic patterns in several species including cocoa (Pokou et al. 2009), coffee (Leroy et al. 1993), eucalyptus (Baudouin et al. 1997), faba beans (*Vicia faba* L.) (Link et al. 1996), oil palm (Bakoumé et al. 2010), switchgrass (*Panicum virgatum* L.) (Martinez-Reyna and Vogel 2008), and timothy (Tanaka et al. 2011).

While fairly detailed studies have been published on genetic diversity and population structure within specific *Brachiaria* species (*B. brizantha* (Vigna et al. 2011), *B. humidicola* (Jungmann et al. 2010), and *B. ruziziensis* (Azevedo et al. 2011)), the phylogenetic relationships between different species in the *Brachiaria* agamic complex are still poorly understood. Germplasm accessions have been assigned to *Brachiaria* species solely on the basis of their morphological characters, particularly the size and shape of racemes (Renvoize et al. 1996). Phylogenies of fewer than 30 accessions from different *Brachiaria* species have been constructed using RAPDs (Ambiel et al. 2008) and internal transcribed spacers (ITS) regions of nuclear ribosomal DNA (Torres Gonzalez and Morton 2005), but the results of these studies are inconsistent and the unavailability of SSRs that amplify across multiple species has so far stalled the construction of more detailed phylogenies.

At first glance, the development of *B. brizantha* and *B. decumbens* heterotic groups seems intuitive. These species are presumably genetically distant, inter-fertile, and possess complementary traits of interest. *B. brizantha* accessions are highly productive, responsive to fertilizer application, and resistant to spittlebug damage, but require moderately fertile soils to thrive and persist under grazing. In contrast, *B. decumbens* accessions are persistent under intensive grazing and tolerant of aluminum toxicity and low-fertility soils, but very susceptible to spittlebug (Rao et al. 1996). However, *Brachiaria brizantha* and *B. decumbens* should not be considered appropriate heterotic groups in the proposed RFRS scheme without first obtaining better estimates of the genetic distance between species. Genetic distance alone is insufficient as a predictor of heterotic patterns, as heterosis has been shown to increase with greater genetic distance between the parents up to a certain optimum distance when it begins to decline (Moll et al. 1965). Therefore, it would also be advisable to evaluate the performance of progeny produced from crosses of parents from the same and opposing genetic groups. This process would be somewhat complicated by the difficulty in achieving recombination between apomictic genotypes, but sexual genotypes from the recurrent selection population could be crossed to genetically close and distant germplasm accessions in order to examine existing heterotic patterns.

Meiotic abnormalities and poor seed set are common even in intraspecific crosses of polyploid *Brachiaria* species, and may be a bigger problem in interspecific hybrids (Mendes-Vieira et al. 2005). Seed set is variable in interspecific crosses between sexual plants from the CIAT recurrent selection population and *B. decumbens* cv. Basilisk, but usually falls within a commercially acceptable range. However, evidence suggests that Basilisk may actually be more closely related to *B. brizantha* than *B. decumbens*. While originally described as *B. decumbens* based on morphology (Oram 1990), Renvoize et al. (Renvoize et al. 1996) re-classified Basilisk as an accession of *B. brizantha* also on the basis of morphological data. A phylogeny constructed using RAPDs (Ambiel et al. 2008) and unusual isozyme banding patterns in Basilisk (CM Meghji personal communication) also support the hypothesis that Basilisk is more accurately classified as *B. brizantha.* A better understanding of the genetic distance between Basilisk and other *B. brizantha* and *B. decumbens* genotypes is necessary to build upon the heterotic effects accumulated over cycles of RS-SCA while developing two populations segregating for reproductive mode.

3.5 Conclusion

Recurrent selection for specific combining ability is an efficient and logistically simple means to exploit heterosis in a fully sexual *Brachiara* breeding population. However, the availability of a reliable molecular marker for apomixis makes RFRS a more efficient method for improving traits of interest in a *Brachiaria* breeding population today. While the utilization of open-pollinated crossing blocks (as opposed to paired-crosses) and requirement of a single marker screening step per cycle still make RS-SCA simpler logistically and slightly less expensive, the proposed RFRS scheme presents two major advantages over the current RS-SCA procedures:

1) Selection will be more effective at loci for which there is partial dominance
2) Complete pedigree information will be available for each genotype evaluated in field trials, facilitating BLUP-based selection of superior hybrid clones as well as apomictic and sexual parents to recombine for the next cycle.

The implementation of other forms of MAS will be very economical in the proposed breeding scheme as all clones in the breeding pipeline, including parents from each heterotic pool as well as all hybrid progeny, will be routinely genotyped for reproductive mode as seedlings. The increasing availability and decreasing cost of next generation sequencing and bioinformatics tools will enable the construction of well-saturated molecular maps of *Brachiaria* in the next few years. The identification of molecular markers linked to quantitative trait loci (QTL) controlling traits of interest in the breeding program, including resistance to biotic and abiotic stresses, should follow shortly thereafter. The additional costs of enriching the breeding populations by screening for the presence of QTL should be very low as DNA extraction and screening with a marker linked to the ASGR locus will already be a routine breeding procedure in the proposed RFRS scheme.

The identification of heterotic groups based on genetic distance and hybrid progeny performance remains an important prerequisite for the initiation of RFRS, but apomixis no longer presents a barrier or obstacle to *Brachiaria* breeding. Using RFRS, breeders can manipulate apomixis effectively and directly exploit accumulated heterotic effects in the apomictic cultivars developed in the *Brachiaria* breeding population and other aposporous species.

References

Ambiel AC, Guaberto LM, Vanderlei TM, Machado Neto NB (2008) RAPD grouping of accessions and cultivars of three *Brachiaria* species. Acta Scientiarum-Agron 30:457–464

Asker SE, Jerling L (1992) Apomixis in plants. CRC Press, Boca Raton

Azevedo ALS, Costa PP, Machado MA, Paula CMP de, Sobrinho FS (2011) High degree of genetic diversity among genotypes of the forage grass *Brachiaria ruziziensis* (Poaceae) detected with ISSR markers. Genet Mol Res 10:3530–3538

Bakoumé C, Galdima M, Tengoua FF (2010) Experimental modification of reciprocal recurrent selection in oil palm breeding in Cameroon. Euphytica 171:235–240

Barrios SCL, Valle CB do, Alves GF, Simeão RM, Jank L (2013) Reciprocal recurrent selection in the breeding of *Brachiaria decumbens*. In: Michalk DL, Millar GD, Badgery WB, Broadfoot KM (eds) Proceedings of the 22nd International Grasslands Congress. 2013 Sept 15–19; Sydney, Australia, p 209

Baudouin L, Baril C, Clement Demange A, Leroy T, Paulin D (1997) Recurrent selection of tropical tree crops. Euphytica 96:101–114

Cheres MT, Miller JF, Crane JR, Knapp SJ (2000) Genetic distance as a predictor of heterosis and hybrid performance within and between heterotic groups in sunflower. Theor Appl Genet 100:889–894

Comstock RE, Robinson HF, Harvey PH (1949) A breeding procedure designed to make maximum use of both general and specific combining ability. Agron J 41:360–367

Grüneberg W, Mwanga R, Andrade M, Espinoza J (2009) Selection methods. Part 5: breeding clonally propagated crops. In: Ceccarelli S, Guimarães EP, Weltzien E (eds) Plant breeding and farmer participation. Food and Agriculture Organization of the United Nations (FAO), Rome, pp 275–322

Hallauer AR, Eberhart SA (1970) Reciprocal full-sib selection. Crop Sci 10:315–316.

Hull FH (1945) Recurrent selection for specific combining ability in corn. J Amer Soc Agron 37:134–145.

Jank L, Barrios SC, Valle CB do, Simeão RM, Alves GF (2014) The value of improved pastures to Brazilian beef production. Crop Pasture Sci. doi:10.1071/CP13319. Epub 2014 Mar 11

Jungmann L, Vigna BBZ, Boldrini KR, Sousa ACB, Valle CB do, Resende RMS et al (2010) Genetic diversity and population structure analysis of the tropical pasture grass *Brachiaria humidicola* based on microsatellites, cytogenetics, morphological traits, and geographical origin. Genome 53:698–709

Khan ZR, Midega CAO, Pittchar JO, Murage AW, Birkett MA, Bruce JA et al (2014) Achieving food security for one million sub-Saharan African poor through push—pull innovation by 2020. Philos Trans R Soc [Biol] 369:20120284

Labate JA, Lamkey KR, Lee M, Woodman WL (1997) Molecular genetic diversity after reciprocal recurrent selection in BSSS and BSCB1 maize populations. Crop Sci 37:416–423

Leroy T, Montagnon C, Charrier A, Eskes AB (1993) Reciprocal recurrent selection applied to *Coffea canephora* Pierre.1. Characterization and evaluation of breeding populations and value of intergroup hybrids. Euphytica 67:113–125

Link W, Schill B, Barbera AC, Cubrero JI, Filippetti A, Stringi L et al (1996) Comparison of intra- and inter-pool crosses in faba beans (*Vicia faba* L).1. Hybrid performance and heterosis in Mediterranean and German environments. Plant Breeding 115:352–360

Martinez-Reyna JM, Vogel KP (2008) Heterosis in switchgrass: spaced plants. Crop Sci 48:1313–1320.

Melchinger AE, Boppenmaier J, Dhillon BS, Pollmer WG, Herrmann RG (1992) Genetic diversity for RFLPs in European maize inbreds 2. Relation to performance of hybrids within versus between heterotic groups for forage traits. Theor Appl Genet 84:672–681

Mendes-Vieira D, Boldrini KR, Mendes-Bonato AB, Pagliarini MS, Valle CB do (2005) Abnormal meiotic behavior in *Brachiaria brizantha* (Poaceae) leading to microspore degeneration. Caryologia 58:396–402

Miles JW (2007) Apomixis for cultivar development in tropical forage grasses. Crop Sci 47:S238–249.

Miles JW, Valle CB do (1996) Manipulation of apomixis in *Brachiaria* breeding. In: Miles JW, Maass BL, Valle CB do (eds). *Brachiaria*: biology, agronomy, and improvement. Cali, Colombia: CIAT and Campo Grande. CNPGC/EMBRAPA, Brazil, pp 164–177.

Miles JW, Valle CB do, Rao IM, Euclides VPB (2004) Brachiariagrasses. In: Moser LE et al. (eds) Warm-season (C_4) grasses. Agron Monogr. 45. ASA, CSSA, SSSA, Madison, pp 745–783

Miles JW, Cardona C, Sotelo G (2006) Recurrent selection in a synthetic Brachiariagrass population improves resistance to three spittlebug species. Crop Sc. 46:1088–1093

Moll RH, Lonnquist JH, Fortuno JV, Johnson EC (1965) Relationship of heterosis and genetic divergence in maize. Genetics 52:139–144

Mutimura M, Everson TM (2011) Assessment of livestock feed resource-use patterns in low rainfall and aluminium toxicity prone areas of Rwanda. Afr J Agric Res 6:3461–3469

Ngendahayo M, Coppens d'Eeckenbrugge G, Louant BP (1988) Self-incompatibility studies in *Brachiaria ruziziensis* Germain et Evrard, *Brachiaria decumbens* Stapf and *Brachiaria brizantha* (Hochst) Stapf and their interspecific hybrids. Phytomorphology 38(1):47–51.

Oram RN (1990) *Brachiaria*. In: Register of Australian herbage plant cultivars. 3rd ed. East Melbourne, Vic., Australia: Australian Herbage Plant Registration Authority, Div. Plant Industry, CSIRO. pp 89–92

Ozias-Akins P, Dik PJ van (2007) Mendelian genetics of apomixis in plants. Ann Rev Gen 41:509–537

Pedraza Garcia FP (1995) Hacia la localización del gen de apomixis en *Brachiaria* usando marcadores moleculares RAPD [thesis] Palmira, Valle del Cauca: Universidad Nacional de Colombia
Piepho HP, Möhring J, Melchinger AE, Büsche A (2008) BLUP for phenotypic selection in plant breeding and variety testing. Euphytica 161:209–228
Pokou ND, N'Goran JAK, Lachenaud P, Eskes AB, Montamayor JC, Schnell R et al (2009) Recurrent selection of cocoa populations in Cote d'Ivoire: comparative genetic diversity between the first and second cycles. Plant Breeding 128:514–520.
Rao IM, Kerridge PC, Macedo MCM (1996) Nutritional requirements of *Brachiaria* and adaptation to acid soils. In: Miles JW, Maass BL, Valle CB do (eds) *Brachiaria*: biology, agronomy, and improvement. Cali, Colombia: CIAT and Campo Grande, Brazil: CNPGC/EMBRAPA, p 53–71
Renvoize SA, Clayton WD, Kabuye CHS (1996) Morphology, taxonomy, and natural distribution of *Brachiaria* (Trin.) Griseb. In: Miles JW, Maass BL, Valle CB do (eds) *Brachiaria*: biology, agronomy, and improvement. Cali, Colombia: CIAT and Campo Grande, Brazil: CNPGC/EMBRAPA, pp 1–15
Savidan YH (1983) Genetics and utilization of apomixis for the improvement of guineagrass (*Panicum maximum* Jacq.). In: Smith JA, Hays VW (eds) Proceedings of the 14th International Grasslands Congress; 1981 Jun 15–24; Lexington, KY. Westview Press, Boulder, pp 182–184
Silva PIT, Martins AM, Gouvea EG, Pessoa M, Ferreira ME (2013) Development and validation of microsatellite markers for *Brachiaria ruziziensis* obtained by partial genome assembly of Illumina single-end reads. BMC Genomics 14:17
Stür WW, Phensavanh P, Gabunada F, Horne P, Khanh TT, Phimphachanhvongsod V et al (2006) A survey of adoption of improved forages in Southeast Asia. Tropical Grasses and Legumes: optimizing Genetic Diversity for Multipurpose Use: Project IP-5: Annual Report 2006. Cali, Colombia: CIAT. pp 129–135
Swenne A, Louant BP, Dujardin M (1981) Induction par la colchicine de formes autotétraploïdes chez *Brachiaria ruziziensis* Germain et Evrard (Graminée). Agron Trop 36:134–141
Tanaka T, Tamaki H, Ashikaga K, Fujii H, Yamada T (2011) Proposal for shift to reciprocal recurrent selections in "clone and strain synthesis" Timothy breeding using molecular marker diversity. Crop Sci 51:2589–2596
Torres Gonzalez AM, Morton CM (2005) Molecular and morphological phylogenetic analysis of *Brachiaria* and *Urochloa* (Poaceae). Mol Phylogenet Evol 37:36–44
Valle CB do, Savidan YH (1996) Genetics, cytogenetics, and reproductive biology of *Brachiaria*. In: Miles JW, Maass BL, Valle CB do (eds) *Brachiaria*: biology, agronomy, and improvement. Cali, Colombia: CIAT and Campo Grande, CNPGC/EMBRAPA, Brazil, pp 147–163.
Viana JMS, Mundim GB, Delima RO, Silva FFE, Resende MDV (2014) Best linear unbiased prediction for genetic evaluation in reciprocal recurrent selection with popcorn populations. J Agric Sci 152:428–438.
Vigna BBZ, Jungmann L, Francisco PM, Zucchi MI, Valle CB do, de Souza AP (2011) Genetic diversity and population structure of the *Brachiaria brizantha* germplasm. Trop Plant Biol 4:157–169
Young BA, Sherwood RT, Bashaw EC (1979) Cleared-pistil and thick-sectioning techniques for detecting aposporous apomixis in grasses. Can J Bot 57:1668–1672

Chapter 4
A New Red Clover 2,4-D-Resistant Cultivar to Improve Broadleaf Weed Control and Elucidate the Molecular Mechanism of Resistance

Patricio Ricardo Munoz, K. H. Quesenberry, A. R. Blount, J. A. Ferrell and J. C. Dubeux

4.1 Introduction

Trifolium pratense L. (red clover) is an important forage legume worldwide, grown alone or in mixture with grasses, second only to *Medicago sativa* L. (alfalfa). It is a high-quality forage and very versatile as it can be grazed, harvested for green chop, made into hay, or lately also used in haylage. The University of Florida Forage Breeding Program has already released high-yielding nondormant cultivars (i.e., Cherokee) (Quesenberry et al. 1993), focusing on resistance to root-knot nematodes (i.e., Southern Belle) (Quesenberry et al. 2005), and mid-dormant yield with RKN resistance (i.e., Barduro) (Quesenberry and Blount 2012). However, the establishment of red clover (*Trifolium pratense* L.) is still a challenge due to factors such as limited rainfall, warm temperatures, but mainly by competition from broad-leaf weeds due to its long seedling stage. While broad-leaf weed control in grasses can be accomplished inexpensively using herbicides like 2,4-Dichlorophenoxyacetic acid (2,4-D), no alternatives exist for red clover, as they are susceptible to this herbicide.

Herbicides were first introduced in US agriculture in 1940s and became an important tool for weed control. The herbicide 2,4-D, introduced in the 1940s, is the third most widely used herbicide in the USA (Prince et al. 2012), with an estimated use of 46 million pounds per year (Gervais et al. 2008), and is one of the five most widely used around the world (Wright et al. 2010). The major use is in natural and cultivated grasslands for forage grazing and hay (U.S. Geological Survey 2009), which represent the fourth largest acreage in the USA.

P. R. Munoz (✉) · K. H. Quesenberry · J. A. Ferrell
Agronomy Department, Institute of Food and Agricultural Sciences, University of Florida, Gainesville, FL 32611, USA
e-mail: p.munoz@ufl.edu

A. R. Blount · J. C. Dubeux
North Florida Research & Education Center, Institute of Food and Agricultural Sciences, University of Florida, Mariana, FL 32446, USA

H. Budak, G. Spangenberg (eds.), *Molecular Breeding of Forage and Turf*,
DOI 10.1007/978-3-319-08714-6_4

Herbicides based on 2,4-D operate by causing uncontrolled cell division in vascular tissue (Gervais et al. 2008). In susceptible plants, this abnormality leads to reduced root uptake of water and nutrients, as well as gradual loss of photosynthesis, vigor, and final death of the plant (Bovey 1971). Work conducted in callus tissue culture indicated that resistance to 2,4-D appears to be related to the ability of cells to accomplish detoxification by conversion to water-soluble metabolites (Bristol et al. 1977). Thus, specific metabolic pathways and mechanisms have been recognized as possibly involved in detoxification, including side-chain degradation, conjugation with amino acids, conjugation with sugars, ring hydroxylation followed by conjugation by sugars among others. However, an increase in the surface wax in tolerant strains also suggested that absorption barrier may be one of the causes of resistance (Devine et al. 1975).

In red clover, the narrow-sense heritability of tolerance to 2,4-D was estimated at nearly 0.50 (Taylor et al. 1989). Studies of inheritance of resistance to other 2,4-D related herbicides (Dicamba) on Kochia and wild mustard have indicated that this resistance is likely conferred by a single dominant allele (Jasieniuk et al. 1996; Preston et al. 2009). In the selfing and short-lived species *Arabidopsis thaliana*, 2,4-D resistance appears to be controlled by a single recessive gene, which also has negative effects in plant growth (Estelle and Somerville 1987).

A draft of the genome of the red clover was recently released (Istvánek et al. 2014) providing the foundation for genome-wide analysis of genetic variation for herbicide resistance. The draft genome sequence comprises an assembly of ~314.6 Mbp (of an estimated 418 Mbp genome size), and 47,398 protein-coding genes were annotated. While this resource is only the first step toward full characterization of the genome of red clover, it provides the data that will be necessary for implementation of methods of genotyping based on sequence-capture (Gnirke et al. 2009) and next generation sequencing. These methods are based on oligonucleotide hybridization for recovery of target regions in the genome, sequencing, and genotyping.

Even though the use of 2,4-D has been widespread for more than 60 years, fewer resistant weeds have been identified (Wright et al. 2010) compared to glyphosate (Heap 2014). Glyphosate resistant weeds have increased mainly due to high selection pressure after the intensive (increase of more than fivefold) and extensive increase in the use of glyphosate since the mid-1990s with the release of glyphosate resistant varieties (Roundup Ready®). Nevertheless, similar challenges to the ones created by the wide use of glyphosate are expected to arise with the future commercialization and adoption of transgenic 2,4-D-resistant crops. This study not only presents the development of a new red clover 2,4-D resistant variety but also describes the plans for this new variety to elucidate the genetic architecture of the resistance traits, the gene(s) controlling the underlying mechanism of resistance, and the genetic process by which the resistance is acquired.

4.2 Methods

4.2.1 Cultivar Development

The selection program that resulted in the development of FL24D red clover was initiated in spring 1985 using a broad-based germplasm bulk of an equal mixture of the cultivars Kenstar, Nolins, and Cherokee. Approximately 20,000 seeds were broadcast seeded into metal flats containing an equal mixture of Arredondo fine sand topsoil and Metro Mix. When seedlings were about 4-week old (3 trifoliate leaf stage) they were sprayed with 1.1 kg a.i. ha^{-1} 2,4-D and re-sprayed using similar rates 3 weeks later. After 1 month, 385 (1.95 %) plants were selected for intercrossing based on survival and regrowth. These plants were transplanted in the field, covered with a cage, and intercrossed with honeybees. Only 63 of the 385 selected plants produced sufficient seed at harvest to use in the next cycle and most of the remaining plants died before flowering. Due to the poor performance of the sprayed plants, the breeding procedure was changed for the next five cycles to phenotypic recurrent selection of half-sib families. Sprayed plants were assigned a visual rating on a 1–5 scale (5 = dead) based on the amount of stunting, stem curvature, and root damage. Families were selected on the basis of mean performance. Ten unsprayed remnant plants from each selected family were taken to the field and intercrossed in isolation by natural pollinators and a hive of honeybees, thus ensuring seed production in each cycle. The first four selection cycles were conducted as part of the Ph.D. research of Dr. Steven Taylor. More details of the experiment can be found in Taylor et al. (1989). Two additional cycles were conducted, but records of exact selection percentages in each cycle were not kept, for the development of the cultivar designated FL24D.

4.2.2 Forage and 2,4-D Evaluation

Variety field experiments were established in the cool season of 2009 in Gainesville, Florida to compare the yield production of the new cultivar with known commercial cultivars. The trial was established in a four replicates randomized complete block design (RCBD) with 2 x 4 m plot size including FL24D and eight commercial cultivars. Plots were harvested twice in the season to measure dry matter yield (kg/ha) on April 15 2010 and May 20 2010. Trial was managed accordingly to University of Florida, Institute of Food and Agriculture Science recommendations (http://edis.ifas.ufl.edu/pdffiles/AA/AA19000.pdf).

Two experiments, one in the greenhouse (2012) and one in the field (2013), were conducted to evaluate the resistance of FL24D to the herbicide 2,4-D. Experiments were established in a split-plot design with three replications with 2,4-D rate (0, 1/2×, 1×, and 2×, where × corresponded to the recommended rate of 1.1 kg a.i./ha) as the whole plot treatment and cultivar FL24D or Southern Belle (as standard susceptible control) as the split-plot treatment. Plants were sprayed with 2,4-D at the trifoliate leaf stage and damage by herbicide was measured at 10 and 17 days after spraying (DAS) in the greenhouse and at 14 and 28 DAS in the field. Plants were rated on a 1–9 scale for damage (9=no damage and 1 dead).

Two experiments (Gainesville and Marianna, Florida) were established in the cool season of 2013–2014 with the objective of measuring the effect of 2,4-D spraying on the yield of FL24D. Experiments were established in a split-plot design with four replicates. The whole-plot treatment was 2,4-D rate (0, 1/2×, and 1×) and the split-plot treatment was cultivars (FL24D, Southern Belle, Barduro, and Freedom). Experiments were sprayed at the trifoliate stage and its yield evaluated in April 2014.

4.3 Results

4.3.1 Cultivar Development

The evaluation of the first four cycles by Taylor et al. (1989) showed that not only the rate required to caused 50% injury in the plant (ID50) increased two- to three-fold after four selection generations (Fig. 4.1) but also that the maximum level of injury at 1× and 4× the 2,4-D label rate (1.12 and 4.5 kg a.i./ha, respectively) consistently decreased after each generation (Fig. 4.2). Additionally, because no plateau in resistance level has been observed, it was likely that continuing with the recurrent selection will increase 2,4-D resistance even further. Two additional cycles were conducted, but records of exact selection percentages in each cycle were not kept. Seed of the population resulting from six cycles of selection for 2,4-D tolerance was designated FL24D, increased in Gainesville, FL in summer 1995 and used for the experiments described hereafter.

4.3.2 Forage and 2,4-D Evaluation

As shown in Table 4.1 dry matter of FL24D was not significantly different than top-producing cultivars developed at the University of Florida (Barduro, Cherokee, and Southern Belle), however, they were all significantly different than the non-Florida cultivars at the first harvest. The second harvest was significantly smaller, and the cultivar Marathon was the top producer. Overall for the season, both FL24D and Barduro were the top producing cultivars with 4800 and 4780 kg/ha, respectively.

Table 4.2 shows the resistance of FL24D compared with the cultivar Southern Belle (SB) in a greenhouse (GH) and field experiments (FD). Higher damage at the

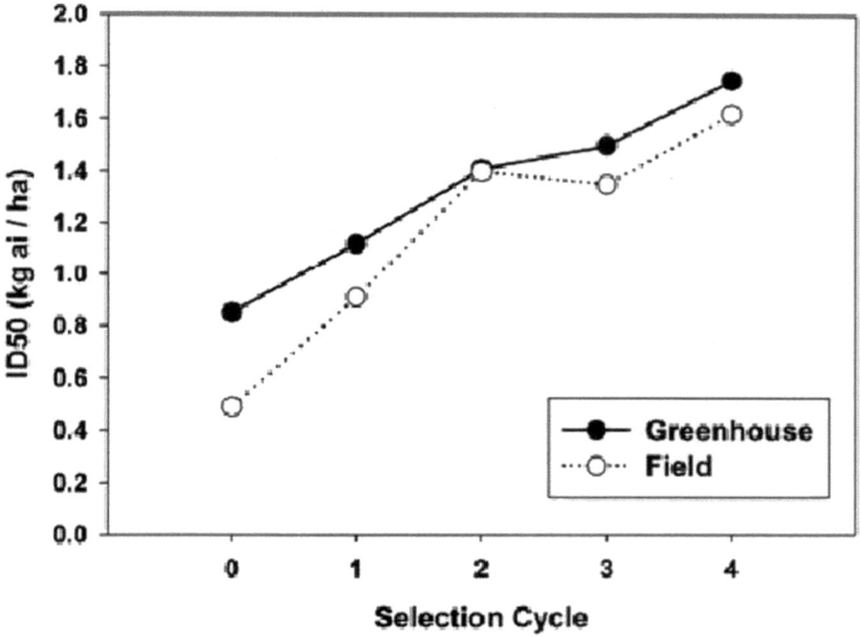

Fig. 4.1 ID50 in response to 2,4-D applications under greenhouse and field conditions in red clover plants from different generations of phenotypic recurrent selection for 2,4-D resistance

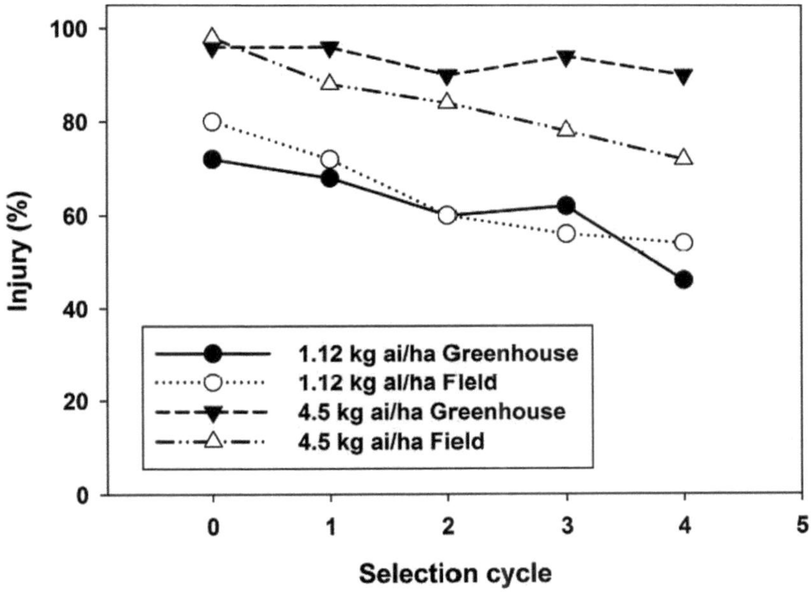

Fig. 4.2 Maximum injury observed after 2,4-D applications at 1× and 4× label rate (1.12 kg a.i./ha) under greenhouse and field conditions in red clover plants from the four first generations of phenotypic recurrent selection for 2,4-D resistance

Table 4.1 Dry matter yield (kg ha^{-1}) of red clover at Gainesville, FL 2009–2010

Entry	Harvest1 04/15/2010	Harvest2 05/20/2010	Total harvest
Barduro	3790 a	1117 bc	4800 a
Southern Belle	3680 a	1065 bc	4750 ab
Cherokee	3530 ab	1040 bc	4570 abc
FL24D	3460 ab	1320 bc	4780 a
Morningstar	3060 bc	820 cd	3880 cd
Freedom	2560 cd	1058 bc	3615 d
Kenton	2530 d	1430 b	3960 bcd
Redland max	2300 d	1980 a	4271 abcd
Marathon	2240 d	1390 b	3630 d
Mean	3040 bc	1157 bc	4195 abcd

Means within a column followed by the same letter are not significantly different, Duncan's multiple range test, $\alpha=0.05$

Table 4.2 Greenhouse (GH) and field (FD) response of FL24D and Southern Belle (SB) to different rates of 2.4-D

Rate	GH 2012 10 days		GH 2012 17 days		FD 2013 14 days		FD 2013 28 days	
	FL24D	SB	FL24D	SB	FL24D	SB	FL24D	SB
2×	4.4a	1.4a	4.9a	1.1a	5.5a	3.3a	6.2a	1.3a
1×	6.5b	1.9a	7.0b	1.2a	7.3b	4.9b	6.6a	1.9a
1/2×	6.4b	2.4a	6.7b	1.5a	7.7b	5.4b	7.4b	4.1b
0x[a]	8.8c	8.3b	9.0c	8.7b	9.0c	9.0c	9.0c	9.0c

Rated on a scale of 1 to 9 where 9 = no damage and 1 = dead. Means within a column followed by the same letter are not significantly different, Duncan's multiple range test, $\alpha=0.05$

[a] damage due to herbicide drift occurred on the GH experiment and explain values below 9 on the 0× application rate

same rated of 2,4-D was observed in the GH when compared to the FD experiment. Damage due by herbicide drift was observed in the GH control treatment (no 2,4-D spraying), explaining the scores below 9. FL24D was significantly less damaged than SB at the different rates of herbicide applied. At the recommended rate of application (1 ×) FL24D has in average a score of 6.9 (across all evaluations, 9 = no damage), while SB has an average a score of 2.5 (where 1 = dead). When the rate was incremented to 2× almost no plants of SB survived (Fig. 4.3).

No significant differences were found for dry matter on the different 2,4-D rates treatments in the Gainesville location, most likely due to unexpected precipitation after spraying. Overall, FL24D showed the highest yield under the recommended application and in average across treatments it produced 2500 kg/ha approximately (Fig. 4.4).

Fig. 4.3 Effect of spraying 2,4-D (2x=2.24 kg ai/ha) on the cultivar Southern Belle (SB) and FL24D under field conditions

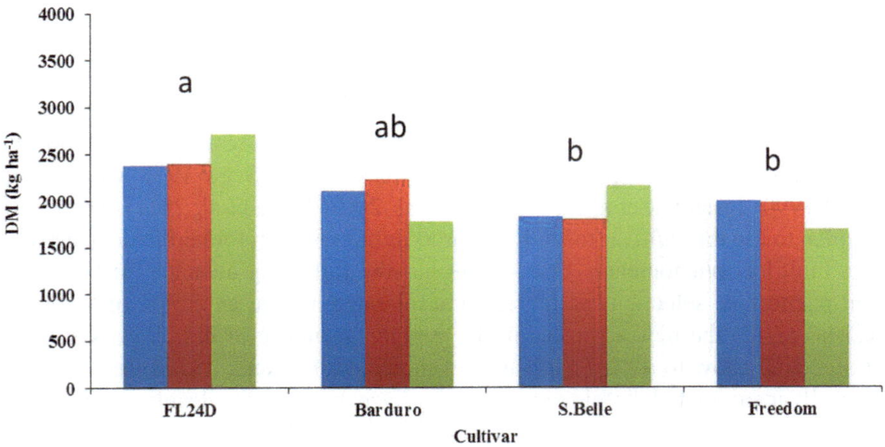

Fig. 4.4 Dry matter for four cultivars averaged across 2,4-D application rates, in Gainesville, Florida, 2014. *Bars* with same letters are not significantly different ($P > 0.05$). (*Blue* = control, *red* = 1/2×, and *green* = 1× of 2,4-D spraying)

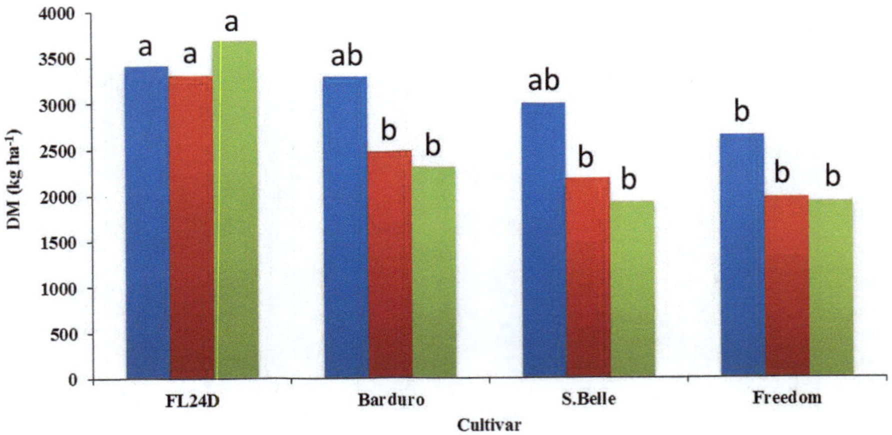

Fig. 4.5 Dry matter for four cultivars and two 2,4-D application rates, in Marianna, Florida 2014. (*Blue* = control, *red* = 1/2×, and *green* = 1× of 2,4-D spraying)

In the Marianna location the 2,4-D treatments showed statistical differences on yield overall (Fig. 4.5). In the susceptible cultivars the yield decreased as the rate of 2,4-D increased, producing the lowest yield with the recommended rate of 2,4-D (1 ×). FL24D was significantly different than all other cultivars under 1/2× and 1× of 2,4-D spraying, producing the highest yield of the experiment under the recommended rate of 2,4-D application. FL24D was not significantly different from Barduro or Southern Belle under the control treatment, but yielded higher than Freedom under no spraying condition.

The almost linear increase in resistance level through the recurrent selection process from generation one to four suggests that the 2,4-D resistance trait might not be the result of a single dominant allele and is more likely to be oligogenic or polygenic. This is especially important because most herbicide resistance evolutionary models are based on single dominant alleles (Gressel and Segel 1978; Gressel and Segel 1990; Jasieniuk et al. 1996; Neve et al. 2003; Neve et al. 2011). Therefore, we plan to use this cultivar to study the process of evolution for herbicide resistance as a multiloci phenomenon. Our system has the following advantages: (1) recurrent phenotypic selection resembles an accelerated natural evolution of herbicide resistance, (2) the new advances in the genome sequence of red clover (Istvánek et al. 2014) allow to design thousands of molecular markers, and (3) red clover is an easily manipulated diploid outcrossing species. Initially, we develop a sequence-capture (*a.k.a.* exome-capture) tool to characterize DNA variation in all 47,398 annotated genes of the red clover genome. Next, we characterize genetic variation and identify changes in allele frequency between bulks of resistant and susceptible plants, in each of the six generations of selection. Finally, we evaluate whether higher 2,4-D herbicide rates accelerate changes in allele frequency in the population

4.4 Discussion

In a series of studies we showed that the yield of the cultivar FL24D is higher or as high as the best cultivars tested in the variety trials. FL24D was selected in six cycles of recurrent selection to have resistance to the herbicide 2,4-D, the resulting cultivar was able to form a normal stand even when twice the recommended rate of 2,4-D was applied. We will develop a high-density of molecular markers to genotype plants from the six generations of selection and thus study both the molecular mechanism involved in the resistance and the genetic process of acquiring resistance (weed resistance evolution).

References

Bovey RW (1971) Hormone-like herbicides in weed control. Econ Bot 25(4):385–400

Bristol DW, Ghanuni AM, Oleson AE (1977) Metabolism of 2,4-dichlorophenoxyacetic acid by wheat cell suspension cultures. J Agric Food Chem 25(6):1308–1314

Devine TE, Seaney RR, Linscott DL, Hagin RD, Brace N (1975) Results of breeding for tolerance to 2,4-D in birdsfoot trefoil1. Crop Sci 15(5):721

Estelle MA, Somerville C (1987) Auxin-resistant mutants of *Arabidopsis* thaliana with an altered morphology. Mol Gen Genet MGG 206(2):200–206

Gervais JA, Luukinen B, Buhl K, Stone D (2008) 2,4-D technical fact sheet; national pesticide information center, Oregon State University Extension Services. http://npic.orst.edu/factsheets/2,4-DTech.pdf. Accessed 02 April 2014

Gnirke A, Melnikov A, Maguire J, Rogov P, LeProust EM, Brockman W et al (2009) Solution hybrid selection with ultra-long oligonucleotides for massively parallel targeted sequencing. Nat Biotechnol 27(2):182–189

Gressel J, Segel LA (1978) The paucity of plants evolving genetic resistance to herbicides: possible reasons and implications. J Theor Biol 75(3):349–371

Gressel J, Segel LA (1990) Modelling the effectiveness of herbicide rotations and mixtures as strategies to delay or preclude resistance. Weed Technol 4(1):186–198

Heap I (2014) The international survey of herbicide resistant weeds. Online. www.weedscience.org. Accessed 04 April 2014

Istvánek J, Jaros M, Krenek A, Řepková J (2014) Genome assembly and annotation for red clover (*Trifolium pratense*; Fabaceae). Am J Bot 101(2):327–337

Jasieniuk M, Brûlé-Babel AL, Morrison IN (1996) The evolution and genetics of herbicide resistance in weeds. Weed Sci 44(1):176–193

Neve P, Diggle AJ, Smith FP, Powles SB (2003) Simulating evolution of glyphosate resistance in *Lolium rigidum* I: population biology of a rare resistance trait. Weed Res 43(6):404–417

Neve P, Norsworthy JK, Smith KL, Zelaya IA (2011) Modelling evolution and management of glyphosate resistance in *Amaranthus palmeri*. Weed Res 51(2):99–112

Preston C, Belles DS, Westra PH, Nissen SJ, Ward SM (2009) Inheritance of resistance to the auxinic herbicide dicamba in kochia (*Kochia scoparia*). Weed Sci 57(1):43–47

Prince JM, Shaw DR, Givens WA, Newman ME, Owen MDK, Weller SC et al (2012) Benchmark study: III. Survey on changing herbicide use patterns in glyphosate-resistant cropping systems. Weed Technol 26(3):536–542

Quesenberry KH, Blount AR (2012) Registration of 'Barduro' mid-dormant red clover. J Plant Regist 2(6):141–145

Quesenberry KH, Prine GM, Ruelke OC, Dunavin LS, Mislevy P (1993) Registration of 'Cherokee' red clover. Crop Sci 33:208–209

Quesenberry KH, Blount AR, Dunavin LS, Mislevy P (2005) Registration of 'Southern Belle' red clover. Crop Sci 45:2123–2124

Taylor SG, Baltensperger DD, Quesenberry KH (1989) Recurrent half-sib family selection for 2,4-D tolerance in red clover. Crop Sci 29(5):1109

U.S. Geological Survey (2009) Estimated agricultural use of glyphosate. http://water.usgs.gov/nawqa/pnsp/usage/maps/compound_listing.php. Accessed 04 April 2014

Wright TR, Shan G, Walsh TA, Lira JM, Cui C, Song P et al (2010) Robust crop resistance to broadleaf and grass herbicides provided by aryloxyalkanoate dioxygenase transgenes. Proc Natl Acad Sci U S A 8:201013154

Chapter 5
Relationship Between Emergence and Genome Content of Iranian Tall Fescue Entries Under Drought Stress

Iman Rohollahi, Toshihiko Yamada, Nayer Azam Khoshkholghsima, Mohsen Kafi and Yoichiro Hoshino

5.1 Introduction

The genus *Festuca* L. is one of the largest in the family Gramineae whose members are widely adapted to a variety of ecogeographical regions (Yamada 2011). *Festuca* is comprised of some 450 species (Clayton and Renvoize 1986) that range from diploid (2n = 2x = 14) to dodecaploid (2n = 12x = 84) in chromosome number (Šmarda and Stančík 2006). Tall fescue is the most economically important forage and turfgrass species in the genus and is grown throughout temperate regions of the world (Saha et al. 2005). Tall fescue is mostly used as a turfgrass in many areas of the world, including parts of North and South America, Europe, and in cooler regions of Asia, Africa, Australia, and New Zealand. The primary reason that Tall fescue does well in cold, arid, and semiarid regions is its ability to survive under drought conditions compared to other cool-season turfgrass species (Fry and Huang 2004). According to Rechinger (Rechinger 1970), Tall fescue is one of the perennial grasses that are endemic to cold and dry regions of central, northeastern, and northwestern of Iran. Cool-season grass species are typically intolerant of drought during the germination stage, thus making turf establishment difficult under conditions of low soil moisture. Seed germination usually requires supplemental irrigation (Christians 2004). Investigation of emergence and seedling performance under conditions of low soil moisture is an efficient process to screen genotypes of cool-season grasses

I. Rohollahi (✉)
Shahed University and university of Tehran, Iran.
e-mail: i.rohollahi@shahed.ac.ir

M. Kafi
Deapertment of Horticulture science, University of Tehran, Karaj, Iran

T. Yamada · Y. Hoshino
Field Science Center for Northern Biosphere, Hokkaido University, Sapporo, Japan

N. A. Khoshkholghsima
Agriculture Biotechnology Research Institute of Iran (ABRII),
P.O. Box 31535-1897, Karaj, Iran

H. Budak, G. Spangenberg (eds.), *Molecular Breeding of Forage and Turf*,
DOI 10.1007/978-3-319-08714-6_5

for drought resistance (Gazanchian et al. 2006). Plant growth parameters such as germination and subsequent seedling growth can be inhibited by soil water deficits (Hsiao and Xu 2000; Mut and Aka 2010). Traits associated with emergence and seedling establishment of cool-season grasses under water stress conditions are rapid germination, enhanced seedling vigor, larger seed size (Mut and Aka 2010), early root initiation, and rapid root extension (Johnson and Asay 1993). Berg and Zeng 2006 and Roohollahi et al. 2009 found that leaf growth decreased and root length increased slightly when exposured to drought stress. Drought plays a significant role in determining emergence rates and seedling development (Berg and Zeng 2006).

In spite of the fact that Tall fescue is relatively drought resistant (Pessarakli 2008), there is considerable genetic variation for this trait among genotypes and populations (Beard and Sifers 1997). Knight and Ackerly (2002) speculated that species with larger genomes size may be more exposed to extinction in extreme environments with shorter growing seasons. Knight et al. (2005) proposed the hypothesis that there is selection in extreme environments against organisms with large genomes. Environmental factors may cause significant selective changes in plant nuclear DNA content (Gregory 2005). Generally, studies looking for associations between genome content, temperature, and precipitation have tended to show both positive and negative correlations depending on the particular species and environment parameters analyzed (Knight et al. 2005). Gregory (2005) reported positive correlations between plant genome size and capacity for growth under cooler conditions. Studies concerning relationships between genome content and the time between seed germination and anthesis provide conflicting results. Several studies have shown positive (Minelli 1996), negative (Ceccarelli et al. 1993), or no significant (Knight and Ackerly 2002) correlation between genome size and generation time or relative growth rate. Correlations between changes in relative genome content and phenotypic traits have previously been observed in species such as *Tall fescue* (Ceccarelli et al. 1993), *Pisum sativum* (Cavallini et al. 1993), and *Helainthus annus* (Natali 1993). Smarda and Bures (2006) confirmed the relationship between DAPI and PI staining in *Festuca* plants on the basis of measurements made in different seasons. DAPI could be used for estimating the absolute DNA content in *Festuca*. Ceccarelli et al. (1992, 1993) examined the collected seeds from 35 and 30 natural populations *F. arundinaceae* from Italian peninsula that germinated on water paper in petridis under sterile condition. Ceccarelli et al. (1993) has reported a significant negative correlation between the seed germination and leaf length with relative genome content of *F. arundinacea* populations.

There is no report on relationships between emergence and seedling establishment under drought conditions and relative genome content of Tall fescue. In this study, we examined how drought stress and relative genome content influences emergence and seedling development in wild entries of Tall fescue during germination.

5.2 Materials and Methods

5.2.1 Plant Materials and Treatments

Sixteen entries were evaluated in the present study. Seeds of 14 wild Tall fescue entries were collected from different cold, arid, and semiarid regions of Iran by the Rangelands and Forestry Research Institute (stored in −20°C for 1 year) (Table 5.1). In addition, two commercial cultivars from Europe ('Bravado' and 'Barleroy') were used as controls (Table 5.1). Each entry was assayed at four levels of soil moisture content 40, 60, 80, and 100% field capacity (FC) condition [approximately −1.4, −0.6, −0.2, and −0.03 matric potentials (MPa), respectively] in a greenhouse study. A factorial experiment was carried out based on a completely randomized design (CRD) with four replicates. Soil volumetric water content was determined by weighing the pots during the experimental period for each treatment. Germination and seedling establishment were monitored for 20 d. Emergence was recorded at the emergence of the leaf above the soil surface of the seedlings in each pot. Days to first emergence (T0) and days required to reach 50% emergence (T50) were recorded as described by Usberti and Valio (1997). The leaf length and the maximum root length for each emerged seedling were measured for each pot at the end of the experiment (20 d after planting). Leaves and roots of each seedling were separated, oven dried (75 °C for 48 h), and weighed. In this study, mean daily germination (MDG) and germination rate (GR) were calculated as described by Hunter et al. (1984) and Maguire (1962), respectively. Seedling vigor index (SVI) was calculated by multiplying the percentage of emergence for each entry by the mean length (cm) of the seedling (root plus leaf) (Abdul-Baki and Anderson 1973).

5.2.2 Estimation Ploidy Level by Nuclear DNA Content Using Flow Cytometry

DNA ploidy levels of each entry were estimated by using DAPI stained flow cytometry (Partec PA; Partec GmbH, Munster, Germany). A total of more 2000 cells were analyzed. DNA content of leaves of *Hordeum vulgare* cv. 'Sultan' ($2n=2x=14$) were used as the internal standard.

5.3 Results

To determine if a relationship between soil moisture content (drought tolerance) and DNA content exists, emergence and seedling growth of the wild Tall fescue entries collected from various sites in Iran were tested under a range of soil moisture conditions.

Table 5.1 The effects of soil moisture content on final emergence, mean daily germination, germination rate, days to 50% emergence, leaf and root length, leaf dry weight, and seedling vigor index of 16 *Tall fescue* entries in seedling stage

		Leaf dry weight		SVI		Root length		Leaf-length		T 50	GR		MDG		FE[a]			
SI 40%	40%	100%	40%	100%	40%	100%	40%	100%	40%	100%	40%	100%	40%	100%[b]	40%	100%	Group †	Entries†
0.53	5.8 abc	11.2 fg	11.8bc	17.8def	5.9 a-d	7.8 cde	8 abcd	10.6 gh	7.6 fg	5.3 fg	8.4 bc	14.3 abc	14.3 abc	4.8 abc	85 ab	96.6 abc	I	Sanandaj
0.50	6.4 ab	13 def	13.0ab	21.7ab	6.1 abc	8.9 ab	8.6 ab	14.3 bc	9.3 e	6 ef	7.2 cd	12.2 b-e	12.2 b-e	4.6 abc	88.3 a	93.3 abc	I	Gonabad
0.50	6.9 a	13.8 def	15.2a	22.3ab	6.4 a	8.9 ab	8.8 a	13.4 bcd	6.6 g	5 g	12.2 a	15.5 a	15.5 a	5 a	100 a	100 a	I	Isfahan
0.34	5.6 a-d	16.9 abc	8.0def	22.6a	6.4 a	8.8 abc	8.2 abc	14.2 bc	15 b	5 g	3.9 fgh	13.7 a- d	13.7 a-d	4.9 ab	55 cd	98.3 ab	II	Semirom
0.35	3 ef	8.7 gh	2.7gh	11.1h	5.6 bcd	7.2 e	3.9 h	11.8 efg	-	9 b	1.6 ijk	6.0 fg	6.0 fg	2.9 e	28.3 fg	58.3 e	III	Borujen
0.37	2.9 ef	8.6 gh	3.7g	14.3g	5.2 de	9 ab	4.6 h	10 h	-	8 c	1.2ijk	8.0 f	8.0 f	3.7 d	35 ef	75 d	III	Kamyaran
0.34	5 cd	14.4 cde	9.0de	20.9abc	5.4 cd	8.2 a-d	7.4 cde	13 cde	12.6 c	6 ef	5.0d- g	13 a-d	13 a-d	4.8abc	70 bc	96.6 abc	II	Mashhad
0.38	5.2 bcd	14.7 cde	8.0def	19.7bcd	6 abc	9.3 a	7.2 de	16 a	11 d	5 g	4.9efg	12.4 b-e	12.4 b-e	3.91 d	60 cd	78.3 d	II	Ardabil
0.30	4.3 de	14.1 cde	7.4ef	20.1abc	5.2 de	8.7 abc	6.2 g	13.6 bcd	14 bc	6.3 de	4.3e- h	10.7 e-h	10.7 e-h	4.5 bc	65 cd	90 bc	II	Karaj
0.09	1.9 fg	11.8 ef	0.5 hi	18.1 cde	1.9 f	8.5 a-d	1.6 i	11.8 efg	-	7 d	0.6 gh	11.2 de	11.2 de	4.7 abc	13.3 gh	93.3 abc	VI	Yasuj
0.38	6.4 ab	17.9 a	6.5 f	22.2 ab	6 abc	8.4 a-d	7 efg	14.5 b	19 a	5 g	2.5 jk	15.5 a	15.5a	4.8 abc	50 de	96.6 bc	II	Barvado
0.34	6.3 abc	17.4 ab	9.8 cde	20.1 a-d	4.6 e	8.3 a-d	6.3 fg	12.5 def	8.3 ef	5 g	6.4 hij	12.8 b-e	12.8 b-e	4.8 abc	90 a	96.6 abc	II	Barleroy
0.19	1.5 g	7.8 h	0.03i	8.8 h	2.3 f	7.6 de	2.3 i	8 i	-	15.3 a	0.1 k	3.9 g	3.9 g	2.8 e	6.7 h	56.66 e	VI	Quchan

Table 5.1 (continued)

		Leaf dry weight		SVI		Root length		Leaf-length		T 50	GR		MDG		FE[a]			
0.49	6.2 abc	12.9 def	7.9 def	16.4 efg	5.3 cde	7.8 cde	7.1 ef	10.8 gh	18.6 a	6 ef	2.9 ghi	11.9 cde	11.9 cde	4.4 c	63.3 cd	88.33 c	II	Tiran
0.40	6.6 a	14.3 cde	12.3 b	20.0 a-d	5.8 a-d	7.9 b-e	7.2 de	12.5 def	5 efg	5 g	9.7 b	14.5 ab	14.5 ab	4.9 ab	95 a	98.33 ab	I	Daran
0.46	6.4 ab	14.7 bcd	7.9 def	16.4 efg	6.1 ab	8.2 b-e	7.8 bcde	11.4 fgh	13.3 c	6 ef	5.2 ghi	11.2 de	11.2 de	3.9 d	65 d	78.33 d	II	Yazdabad

FE final emergence, *MDG* mean daily germination, *GR* germination rate, T50 days required to reach 50 % emergence, *LL* leaf length, *RL* root length, *SVI* seedling vigor index and *LD* leaf dry weight

[a] Classification was conducted on PCA (see Fig. 5.2)

[b] 100 and 40 % Field soil moisture capacity

5.3.1 Emergence Traits

While all of the entries displayed some emergence at 40 % FC, the averages for the entries collected from Quchan and Yasuj were low (Table 5.1). The highest final emergence (100 %) was exhibited by entry from Isfahan, and 90 % for the entry from Barvado at 40 % FC (Table 5.1). Final emergence for the Quchan, Yasuj, Borujen, and Kamyaran entries was 6.7, 13.3, 28.3, and 35 % at 40 % FC, respectively (Table 5.1). Soil water content at field capacity did not provide the greatest final emergence for all entries. For instance, Tall fescue collected from Isfahan, Yazdabad, Semirom, and Ardabil had optimum final emergence, at 80 % FC. Whereas most entries achieved 50 % or greater emergence at 40 % FC, the low soil water content delayed days to 50 % emergence compared to emergence at FC between 0 to 14 days depending on the entry (Table 5.1). A comparison of the two soil water contents (100 and 40 % FC) revealed a significant decrease in GR (60 %) and MDG (60 %) across all entries (Table 5.1). The Isfahan entry showed the maximum MDG (5) and GR (12.22) at 40 % FC (Table 5.1). At 40 % FC, GR in Sanandaj (8.43) and Gonabad (7.25) were higher than all others after the Isfahan entry, whereas the lowest GR obtained from Quchan (0.1) and Yasuj (0.6) (Table 5.1). GR and MDG were 6.4 and 4.5 in 40 % FC for Barvado (Table 5.1). At 40 % FC, Gonabad (14.35), Semirom (14.2), Kamyaran (13.6), and Isfahan (13.4) entries displayed the best MTG, respectively. In contrast, the Quchan (0.1), Yasuj (0.6), and Borujen (1.6) entries were slow to germinate and sensitive to low soil water content (Table 5.1).

5.3.2 Seedling Growth

The highest SVI measurements occurred for the Isfahan and Gonabad entries and the lowest for Quchan and Yasuj with 40 % soil water content (Table 5.1). Under 40 % FC SVI values decreased by 70 % and 51 % in Barvado and Barleroy, respectively (Table 5.1).

Sensitivity indices (SI 40 %) at 40 % FC for the Tiran, Gonabad, and Isfahan entries had the lowest reduction in leaf dry weight (Table 5.1), and root dry weight either increased (Tiran and Isfahan) or was unchanged (Semirom and Daran). Leaf dry weight and root dry weight decreased between 100 and 40 % FC by 64 and 21 % respectively in Barvado and Barleroy. For all entries, the leaf and root length of seedlings decreased significantly by more than 50 and 36 % respectively between 100 and 40 % FC (Table 5.1). Under 40 % FC, Isfahan had the greatest leaf (8.8 cm) and root (6.4 cm) length, in comparison with other genotypes, whereas Yasuj had the least leaf (1.6 cm) and root (1.9 cm) growth (Table 5.1). Leaf length and root length decreased by 52 and 21 % respectively in Bravado while it decreased by 51 and 28 % in Barleroy in 40 % compared to 100 % FC (Table 5.1).

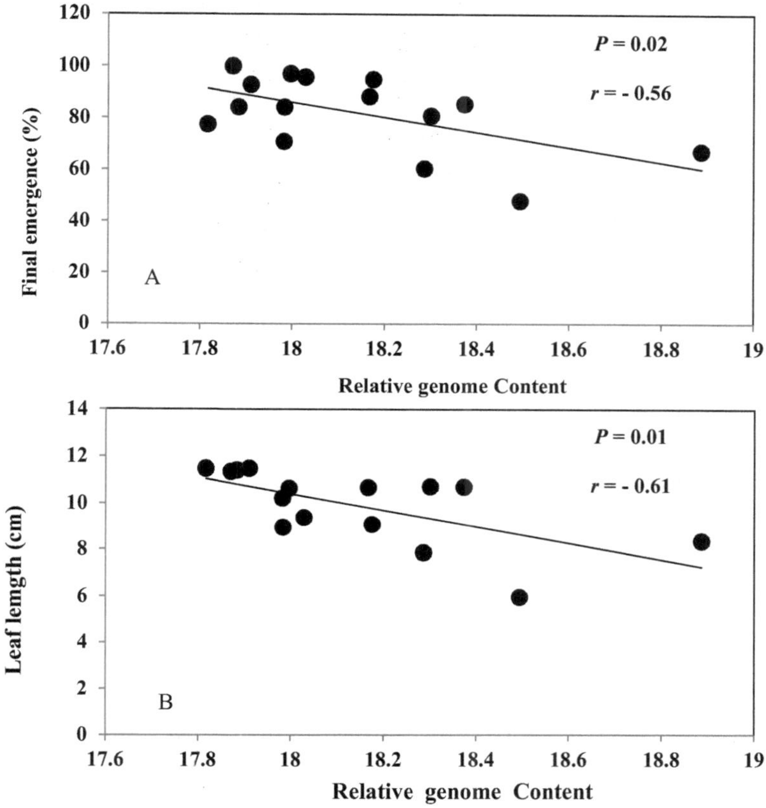

Fig. 5.1 Correlation of the estimated DNA content with the emergence percentage of seeds (**a**) and the leaf length (**b**) in 13 Iranian wild Tall fescue entries and 2 turf cultivars (Barvado and Barleroy) within hexaploid entiries

5.3.3 Estimated DNA Content

Except one entry diploid (Brojen), all entries were hexaploid but more research is needed to confirm this finding. Our results indicate the significant negative correlation between emergence percentage (r=0.56) and leaf length (r=0.61) with relative genome content (Fig. 5.1) within hexaploid entries.

5.4 Discussion

There were substantially genetic differences in emergence and seedling development under conditions of drought stress among the different wild populations of Tall fescue collected at different locations within Iran. The Tall fescue entry

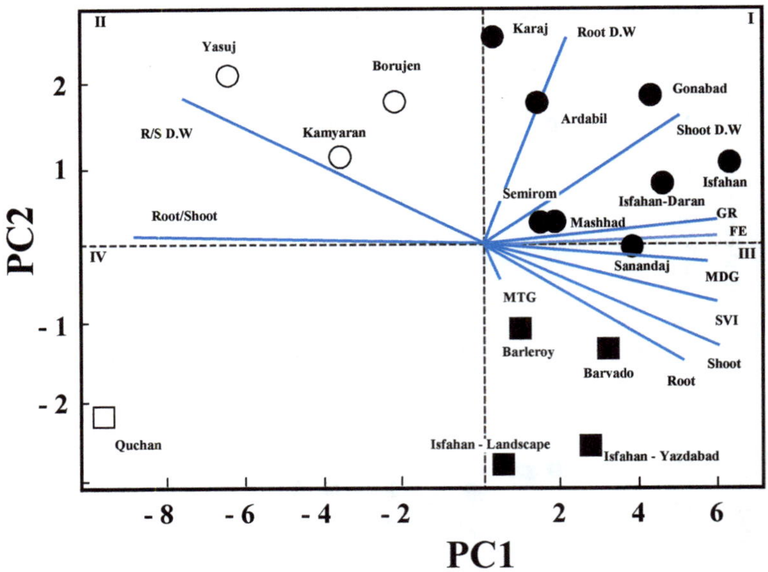

Fig. 5.2 Biplot for diversity of 16 Tall fescue entries at four soil moisture (Fc, 80, 60, and 40% FC) in seedling emergence stage, showing which tolerant entries to low soil moisture had higher values PC1 and PC2. PC1 and PC2 are first and second principal component. *FE* final emergence, *MDG* mean daily germination, *SVI* seedling vigor index, *GR* germination rate, *MTG* mean time to germination, *Shoot* shoot length, *Root* root length, *Root/Shoot* root to shoot ratio length, *Shoot D.W* shoot dry weight, *Root D.W* root dry weight, *R/S D.W* root to shoot ratio dry weight

collected from Isfahan had the greatest final emergence in low soil water (− 1.4 MPa). Additionally, shoot dry weight and root length decreased by 35 and 29% respectively, at 40% FC compared with 100% FC. Several entries were not able to reach 50% emergence at 40% FC. This finding was in agreement with Gazanchian et al. (2006), who found that none of the perennial grasses tested would emerge from soils at 25% FC. In our studying, it was demonstrated that Iranian Tall fescue entry needs a minimum of 40% FC soil water for germination and subsequent seedling establishment.

Principal component analysis (PCA) is a technique for screening multivariate data and is quite useful in the separation of similar experimental units belonging to the same group (Johnson 1998). In our study, the biplot derived from the PCA effectively separated and grouped the Tall fescue entries based on their responses to soil moisture levels (Fig. 5.2). On the basis of PCA biplot analysis (Table 5.2), the Tall fescue entries were grouped into four groups: tolerant (I), intermediate (II and III), and susceptible (IV) in response to low soil moisture at germination and seedling stage (Fig. 5.2).

Quchan and Yasuj entries in group IV, with low final emergence (6.7 and 13.3%), delayed emergence, and a low seedling vigor index at 40% FC (0.03, 0.5 respectively), were the most sensitive entry for germination and emergence under low soil water conditions. The findings of the current study are consistent

Table 5.2 Eigenvectors from the first three principal components axes of the traits used to classify 16 genotypes of *Tall fescue* in seedling stage at four soil moisture levels

Traits	Irrigation conditions	Principal component axis		
		1	2	3
Germination rate	% FC[a]	0.300	−0.023	−0.192
	80 % FC	0.303	−0.061	0.107
	60 % FC	0.318	0.016	−0.150
	40 % FC	0.262	0.107	0.335
Seedling vigor index	% FC†	0.298	−0.125	−0.060
	80 % FC	0.304	−0.142	0.049
	60 % FC	0.316	−0.122	−0.145
	40 % FC	0.295	0.021	−0.239
Final emergence (%)	% FC†	0.279	−0.036	0.377
	80 % FC	0.277	−0.211	0.322
	60 % FC	0.298	0.008	−0.377
	40 % FC	0.294	0.016	0.210
Mean daily germination	% FC†	0.278	−0.053	−0.385
	80 % FC	0.284	−0.194	0.306
	60 % FC	0.300	0.014	−0.364
	40 % FC	0.294	0.011	0.212
Mean time to germination (d)	% FC†	−0.296	−0.002	−0.085
	80 % FC	−0.304	−0.040	0.065
	60 % FC	−0.293	−0.073	−0.193
	40 % FC	0.201	0.150	0.574
Stem length (cm)	% FC†	0.246	−0.157	0.479
	80 % FC	0.244	0.064	−0.406
	60 % FC	0.239	−0.295	−0.100
	40 % FC	0.292	−0.072	0.018
Root length (cm)	% FC†	0.155	−0.228	0.118
	80 % FC	0.176	−0.146	−0.040
	60 % FC	0.035	−0.164	0.470
	40 % FC	0.256	0.018	0.071
Root to shoot length ratio (R/S)	% FC†	−0.234	0.097	−0.476
	80 % FC	−0.215	−0.167	0.513
	60 % FC	−0.257	0.229	−0.127
	40 % FC	−0.297	−0.046	−0.146
Shoot dry weight	% FC†	0.282	0.238	0.144
	80 % FC	0.282	0.277	−0.067
	60 % FC	0.253	0.290	0.245
	40 % FC	0.297	−0.289	−0.026

Table 5.2 (continued)

Traits	Irrigation conditions	Principal component axis		
		1	2	3
Root dry weight	% FC†	0.145	0.609	0.081
	80 % FC	0.084	0.611	0.264
	60 % FC	0.051	0.572	0.105
	40 % FC	0.078	−0.706	0.123
Root/Shoot Dry weight ratio	% FC†	−0.181	0.522	−0.106
	80 % FC	−0.224	0.323	0.375
	60 % FC	−0.209	0.378	−0.205
	40 % FC	−0.275	−0.242	−0.129
Variance explained by each factor	% FC†	67.7	13.6	8.2
	80 % FC	62.1	14.1	10.9
	60 % FC	57.3	18.5	10.3
	40 % FC	68.1	11.9	9.5

[a] Field Capacity

with those of Gazanchian et al. (2006), who found that the Tall fescue entry from Quchan had the lowest tolerance to low soil water for seed germination. A high-correlation coefficient between SVI and the other traits in our work, particularly at 40 % FC, indicated that it could be used as a selection criterion for improving tolerance to low soil water for emergence and seedling development in cool season grasses.

In our study, the effect of low soil water content was more severe on leaf than on root length. As a result, root growth is probably a better indicator of establishment and survival than leaf growth in tall fescue under drought conditions. In cool-season grass species, quick emergence depends on factors associated with the genotypes, the germination environment, and seed characteristics (Gazanchian et al. 2006; Mut and Aka 2010) including seed dormancy, seed reserves, and seed weight (Andrews et al. 2008). It is well known that rapid and complete germination are important during the establishment of grass species (Mut and Aka 2010; Larsen and Bibby 2004). Seedling vigor index and final emergence of the tolerant genotypes were significantly higher under lower soil moisture levels. Studies on identification indices of drought resistance indicated that final emergence, leaf length, root length, and seedling vigor index were most important and better indicators of establishment in tall fescue (*F. arundinaceae*). Our study showed that change in the basic amount of relative genome content within hexaploid Tall fescue entries might have an influence on *Festuca* establishment by affecting seed germination and growth rates during early development under drought stress. Similar to our results on negative correlation between final emergence, leaf growth, and relative genome content has also been observed by Ceccarelli et al. (1992, 1993) in *F. arundinacea*. Thus, the variation in the basic amount of genome content within *Tall fescue* which is distributed in wide geographical area may represent a factor in the evolution of this species.

In conclusion, our work revealed significant variation between Tall fescue entries from different ecological regions of Iran in response to low soil water content. Low soil moistures content at 40% FC decreased the FE, SVI, and seedling characters such as leaf and root length, especially in susceptible entries in comparison with tolerant genotype. Ultimately, on the basis of results and cluster analysis, FE, SVI, leaf and root length were the critical and important traits for the evaluation of drought tolerant Tall fescue entry with appropriate genome content. Wild entries represent valuable germplasm for the development of Tall fescue turfgrass cultivars with improved establishment under conditions of reduced soil moisture.

Acknowledgments This research was supported under Iran National Science Foundation Projects funding scheme (project number 90001304). We thank Iran National Science Foundation for their financial support.

References

Abdul-Baki AA, Anderson JD (1973) Relationship between decarboxylation of glutamic acid and vigor in soybean seed. Crop Sci 13:227–232

Andrews M, Douglas A, Jones AV, Milburn CE, Porter D, Mckenzie BA (2008) Emergence of temperate pasture grasses from different sowing depths: importance of seed weight, coleoptile plus mesocotyl length and shoot strength. Ann Appl Biol 130:549–560

Beard JB, Sifers SI (1997) Genetic diversity in dehydration avoidance and drought resistance within *Cynodon* and *Zoysia* species. Int Turfgrass Soc Res J 8:603–610

Berg LVD, Zeng YJ (2006) Short communicate. Response of South African indigenous grass species to drought stress induced by polyethylene glycol (PEG) 6000. S Afr J Bot 72:284–286

Bor NL (1970) Gramineae-Festuceae. In: Flora Iranica (Ed Rechinger KH), Akademische Druck- u. Verlagsanstalt. Graz 70:450–452 (In German.)

Cavallini A, Natali L, Cionini G, Gennai D (1993) Nuclear DNA variability within *Pisum sativum* (Leguminosae): nucleotypic effects on plant growth. Heredity 70:561–565

Ceccarelli M, Falistocco E, Cionini PG (1992) Variation of genome size and organization with in hexaploid *Festuca arundinacea*. Theor Appl Genet 83:273–278

Ceccarelli M, Minelli S, Falcinelli M, Cionini PG (1993) Genome size and plant development in hexaploid *Festuca arundinacea*. Heredity 71:555–560

Christians N (2004) Fundamental of Turfgrass Management. Published by John Wiley & Sons. Hoboken, NJ. 172–188

Clayton WD, Renvoize SA (1986) Genera Graminum: grasses of the world. Kew Bull, London (H.M.S.O)

Fry J, Huang B (2004) Applied Turfgrass Science and Physiology. Wiley, New Jersey

Gazanchian A, Khosh Kholgh Sima NA, Malboobi MA, Majidi Heravan E (2006) Relationships between Emergence and Soil Water Content for Perennial Cool-Season Grasses Native to Iran. Crop Sci 46:544–553

Gregory RT (2005) The evolution of the genome. Elsevier Academic Press, New York 740 p

Hsiao TC, Xu LK (2000) Sensitivity of growth of roots versus leaves to water stress: biophysical analysis and relation to water transport. J Exp Bot 51:1595–1616

Hunter EA, Glasbey CA, Naylor REL (1984) The analysis of data from germination tests. The J Agr Sci 102:207–213

Johnson DE (1998) Applied multivariate methods for data analysts. Duxbury Press, New York

Johnson DA, Asay KH (1993) Viewpoint: selection for improved drought response in cool-season grasses. J Range Manage 46:194–202
Knight CA, Ackerly DD (2002) Variation in nuclear DNA content across environmental gradients: a quantile regression analysis. Ecolo Lett 5:66–76
Knight CA, Molinari NA, Petrov DA (2005) The large genome constraint hypothesis: evolution, ecology and phenotype. Ann Bot 95:177–190
Larsen SU, Bibby BM (2004) Use of germination curves to describe variation in germination characteristics in three turfgrass species. Crop Sci 44:891–899
Maguire JD (1962) Speed of germination-Aid in selection and evaluation for seedling emergence and vigor. Am Soc Agron 2:176–177
Minelli S, Moscariello P, Ceccarelli M, Cionini PG (1996) Nucleotype and phenotype in *Vicia faba*. Heredity 76:524–530
Mut Z, Aka H (2010) Effect of seed size and drought stress on germination and seedling growth of Avena sativa. L. Bulg J Agri Sci 16:459–467
Natali L, Cavallini A, Cionini G, Sassoli O, Cionini PG, Durante M (1993) Nuclear changes within *Helainthus annuus* L: changes with in single progenies and their relationship with plant development. Theor Appl Genet 85:506–512
Pessarakli M (2008) Hand Book of Turfgrass Management and Physiology. CRC Press. Taylor & Francis publishing company, Florida, 690 p
Roohollahi I, Kafi M, Naderi R (2009) Drought reaction and rooting characteristic in response to plant growth regulators on *Poa pratensis* cv. Barimpala. J Food Agric Environ 8:285–288
Saha MC, Mian R, Zwonitzer JC, Chekhovskiy K, Hopkins AA (2005) An SSR and AFLP based genetic linkage map of tall fescue (*Festuca arundinacea* Schreb. Theor App Genet 110:323–336
Šmarda P, Stančík D (2006) Ploidy level variability in South American fescues (*Festuca* L., Poaceae): use of flow cytometry in up to 5 1/2-year-old caryopses and herbarium specimens. Plant Biol 8:73–80
Smarda P, Bures P (2006) Intraspecific DNA content variability in *Festuca pallens* on different geographical scales and ploidy. Ann Bot 98:665–678
Usberti R, Valio IFM (1997) Osmoconditioning effects on germination of Guinea grass (*Panicum maximum*) seeds. Seed Sci Technol 25:303–310
Yamada T (2011) Festuca. In: Kole C (ed) Wild crop relatives: genomic and breeding resources, millets and grasses, Springer, New York, p 153–164

Chapter 6
Genome-Wide Analysis of AP2/ERF Family Genes in *Lotus corniculatus* Shows *LcERF054* and *LcERF080* Enhance Salt Tolerance

Zhan-Min Sun, Mei-Liang Zhou, Xing-Guo Xiao, Yi-Xiong Tang and Yan-Min Wu

6.1 Introduction

Soil salinity is one of the most important abiotic stress problems that inhibit growth and reduce the potential yields of crop plants. Deciphering the mechanism of plants salt stress response is of critical importance for the development of rational breeding and transgenic strategies leading to salt stress tolerance in plants (Zhou et al. 2010). Apetala2/Ethylene response factor (AR2/ERF) is a large, and one of the most important families of plant-specific transcription factors which binds to some specific *cis*-acting elements for the regulation of many stress-related gene expression (Sakuma et al. 2002; Mizoi et al. 2012). On the basis of the number of AP2/ERF domains encoded and the gene function, the AP2/ERF gene family has been divided into four subfamilies named the AP2 (which contains two AP2 domains), C-repeat binding factor (CBF)/DREB (single AP2 domain, A-subfamily), ERF (single AP2 domain, B-subfamily), and RAV (related to ABI3/VP1, thus have a B3 domain in addition to the AP2 domain) (Sakuma et al. 2002; Nakano et al. 2006).

Different subgroups play multiple roles in plants. AP2 and RAV subfamilies factors were shown to be mainly involved in plant development (Wurschum et al. 2006; Ohto et al. 2009; Ripoll et al. 2011; Dinh et al. 2012; Mishra et al. 2012; Zhao et al. 2012). ERF family proteins contain a single AP2/ERF domain, and are sometimes further divided into two major subfamilies, the CBF/DREB (DRE binding factor) subfamily and the ERF subfamily, respectively (Sakuma et al. 2002). CBF/DREB subfamily proteins play a crucial role in the response of plants to abiotic stresses, such as cold, drought, and salt (Agarwal et al. 2006; Zhou et al. 2012).

Y.-M. Wu (✉) · Z.-M. Sun · M.-L. Zhou · Y.-X. Tang
Biotechnology Research Institute, Chinese Academy of Agricultural Sciences, Beijing 100081, China
e-mail: wuyanmin211@hotmail.com

Z.-M. Sun · X.-G. Xiao
State Key Laboratory of Plant Physiology and Biochemistry, College of Biological Sciences, China Agricultural University, Beijing 100094, China

H. Budak, G. Spangenberg (eds.), *Molecular Breeding of Forage and Turf*,
DOI 10.1007/978-3-319-08714-6_6

The ERF subfamily factors are mainly involved in response to biotic and abiotic stresses, such as pathogenesis (Huh et al. 2012), cold (Tian et al. 2011), drought (Zhang et al. 2013), and salt (Zhai et al. 2013), by recognizing the GCC box (Hao 1998). Recently, a number of studies presented that many *ERF* genes increased the plants salinity tolerance. Overexpression of *MsERF8* and *GmERF7* in tobacco plants, *SlERF5* in tomato, and *OsAP21* in *Arabidopsis* resulted in higher tolerance to salinity than with nontransgenic plants (Zhai et al. 2013; Chen et al. 2012; Pan et al. 2012; Jin et al. 2013). Consequently, the *ERF* genes can be used to produce transgenic plants with higher tolerance to abiotic stresses such as drought and high salt (Zhou et al. 2012).

L. corniculatus L. (Bird's-foot trefoil), a member of perennial *Fabaceae*, is considered the most agriculturally important forage plant (Swanson et al. 1990). Characterization of AP2/ERF family genes in *L. corniculatus* can help us understand the molecular mechanisms of resistance to abiotic stress, and thus facilitate the development of *L. corniculatus* varieties with improved salt or drought tolerance. Until now, only a few members of the ERF subfamily have been investigated in *Lotus* mainly focusing on nodule formation (Sato et al. 2008). To investigate the salt and drought stress response mechanism of *AP2/ERF* genes in *L. corniculatus*, the *L. japonicus* genome DNA database was surveyed and the expression profile of 106 *LcERF* genes under salt stress was also characterized. In addition, the biological and functional analysis of the *LcERF054* and *LcERF080* genes was investigated in transgenic *Arabidopsis* plants.

6.2 Materials and Methods

6.2.1 *Bioinformatic Analysis of* L. corniculatus *AP2/ERF Genes*

The *Arabidopsis* AP2/ERF family protein sequences were extracted from the database of DATF (the database of *Arabidopsis* transcription factors) (Guo et al. 2005). The amino acid sequence of the AP2/ERF domain from *Arabidopsis* AtCBF1 (AT4G25490.1) was used as a query sequence to search against the databases from NCBI and *L. japonicus* genome database (Zhou et al. 2012; Sato et al. 2008) using the BLASTP program at the e-value of 1e-3 to avoid false positives. We also searched the database using full amino acid sequences of other AP2/ERF, as query sequences to confirm that the search had been exhaustive (Zhou et al. 2012). SMART (Letunic et al. 2012) database searches were used to confirm and classify each predicted *LcAP2/ERF* gene. A multiple alignment analysis between *L. japonicus* and *Arabidopsis* AP2/ERF proteins was performed with ClustalX2 (Chenna 2003). Phylogenetic trees were constructed using the MEGA version 5.05 program with the neighbor-joining (NJ) method (Saitou and Nei 1987; Tamura et al. 2011). Conserved motifs were identified from the multiple-alignment analyses using ClustalX2 and MEME (Bailey et al. 2009). Intron/exon structure analysis was analyzed using the GSDS.

6.3 Plant Materials and Treatments

L. corniculatus cv Leo seeds were sterilized and germinated on the Murashige-Skoog (MS) agar medium in a growth chamber with 16 h of light/8 h of dark, 30-day-old seedlings were transferred to the MS medium supplemented with 150 mM NaCl, The seedlings were harvested at 12 h, and 24 h after treatment. Untreated seedlings were used as control. All samples were frozen in liquid nitrogen and stored at −80 °C until use.

6.4 RNA Extraction and Quantitative Real-Time PCR Analysis (qRT-PCR)

Total RNA was isolated from leaves using RNAprep pure Plant Kit (Tiangen Biotech., Beijing, China). About 2 μg of RNA of each sample was reverse transcribed into cDNA using TransScript cDNA synthesis supermix (TransGen Biotech.). ABI Prism 7000 sequence detection system (Applied Biosystems, Carlsbad, CA) was used for qRT-PCR amplification using default parameters. Primers for qRT-PCR are given in Supplementary Tables 2.3 and 2.4. The polyubiquitin gene was used as an internal control. At least three biological and three technical replicates were performed for each reaction. The data was analyzed using $2^{-\Delta\Delta CT}$ method (Livak and Schmittgen 2001).

6.4.1 Subcellular Localization and Trans-activation Assay of LcERF054

To determine its exact subcellular location, *LcERF054/LcERF080* cDNA was fused in-frame with GFP, The entire coding region without the stop codon of the target gene amplified by PCR was inserted into the *Hind*III and *Sal*I sites of vector p163-GFP (Zhou et al. 2012) to generate CaMV35S: *LcERF054/LcERF080*::GFP. Empty vector was used as a control. The recombinant constructs of the *LcERF054/LcERF080*-GFP fusion gene or the GFP gene alone were transformed into *Arabidopsis* protoplasts using polyethylene glycol-mediated method (Yoo et al. 2007), or particle bombardment was performed to introduce the fusion construct into onion epidermal cells (Kinkema et al. 2000). The transformed cells were incubated for at least 16 h in the dark. The transformed protoplasts were examined using a confocal laser scanning microscope (Zeiss LSM700).

The yeast strain AH109 *Saccharomyces cerevisiae*, containing the *His3*, *Ade2* reporter genes with GAL4-binding elements in the promoter, and the vector pDEST32 (Invitrogen, Carlsbad, CA), containing the GAL4 DNA-binding domain (BD) and *Leu2* reporter gene, were used for testing the transactivation of *LcERF054*. The full length coding region of *LcERF054* was cloned into the pDEST32 vector by using the Gateway technology (Invitrogen). The construct plasmids were introduced into the yeast strain AH109 with the reporter gene *His3* and *LacZ* following the manufacturer's instructions (Clontech, Palo Alto, CA). Transformants were plated on Leu^- and $Leu^- His^- Ade^-$ medium, respectively. β-galactosidase filter-lifted assay was performed as described in the Yeast Protocols Handbook (Clontech), The relative quantification of β-galactosidase activity was carried out following the method of Chen et al. (2003). The data for β-galactosidase relative activity were averaged from triplicate samples and three independent experiments.

6.4.2 *Generation of Transgenic* Arabidopsis

The CDS sequence of *LcERF054/080* was amplified by PCR using the gene-specific primers containing attB sites (underlined): 5′-GGG GAC AAG TTT GTA CAA AAA AGC AGG CTG C -3′ and 5′-GGG GAC CAC TTT GTA CAA GAA AGC TGG GT -3′, The PCR product was cloned into the pH7GW2D to generate the 35S:*LcERF054/080* construct. The plasmid was introduced into *Agrobacterium tumefaciens* GV3101 strain. Four-week-old *Arabidopsis* ecotype Columbia plants were transformed by the *Agrobacterium* using the floral dip method (Clough and Bent 1998). Primary transformants were screened on MS agar medium containing 25 mg/L hygromycin and further identified by RT-PCR.

6.4.3 *Determination of Physiological and Biochemical Properties of Transgenic* Arabidopsis

Seeds of both wild-type (WT) and *LcERF054/080* transgenic *Arabidopsis* (T3) were sterilized and germinated on MS medium for 10 days and then transferred to soil. Thirty-day-old, soil grown *LcERF054/080* transgenic plants and the control plants were irrigated with water containing NaCl (100, 200, and 300 mM) at intervals of 4 days for three applications and were monitored during the next weeks. The experiment was repeated three times. The malondialdehyde (MDA), soluble sugar and proline content was determined as described by Duan et al. (2005), Lehner et al. (2006), and Bates et al. 1973, respectively.

6.5 Statistical Analysis

All data were examined using one-way analysis of variance (ANOVA) (General Linear Models on GLM procedure). The Student Neuman–Keuls test was used to identify the means which differed if the ANOVA test indicated significance. *P* value of 0.05 was considered to be significant.

6.6 Results

6.6.1 Bioinformatic Identification of AP2/ERF Family Genes in L. japonicus

A total of 127 full-length AP2/ERF family genes were identified which encode the AP2/ERF domain in *L. japonicus*. Among the AP2/ERF gene families, 19 were assigned to the AP2 family because of containing 2 AP2/ERF domains and homology. One gene was predicted to encode one AP2 domain together with one B3 domain and was assigned to the RAV family. One hundred and seven genes were predicted to encode proteins containing a single AP2/ERF domain. Out of 107, 106 genes were assigned to the ERF family. The remaining one gene includes an AP2/ERF like domain sequence, but its homology appears quite low in comparison with the other AP2/ERF genes. Therefore, this gene was designated as a soloist. Given the above classification, the 106 genes of the ERF family were subjected to further analyses. These genes were further classified into two subfamilies on the basis of similarity of the amino acid sequences. Forty genes were identified as the CBF/DREB subfamily, and 66 genes were predicted to encode the ERF subfamily. SMART analysis indicated that the AP2/ERF domain of each of the 127 *LcAP2/ERF* genes was typical, thereby certifying to their reliability.

To determine the evolutionary relationship among LcERF proteins, an unrooted NJ phylogenetic tree was constructed (Fig. 6.1). The result showed that the 106 LcERF members were divided into 12 subgroups, designated A-1 to A-6 and B-1 to B-6, in accordance with the classification described by Sakuma et al. (2002). Subgroups A-1 to A-6 represent the CBF/DREB subfamily, and subgroups B-1 to B-6, the ERF subfamily. Comparative analyses of the phylogenetic tree suggested that the classification of the *L. japonicas* ERF family was similar and applicable to the *Arabidopsis* ERF family.

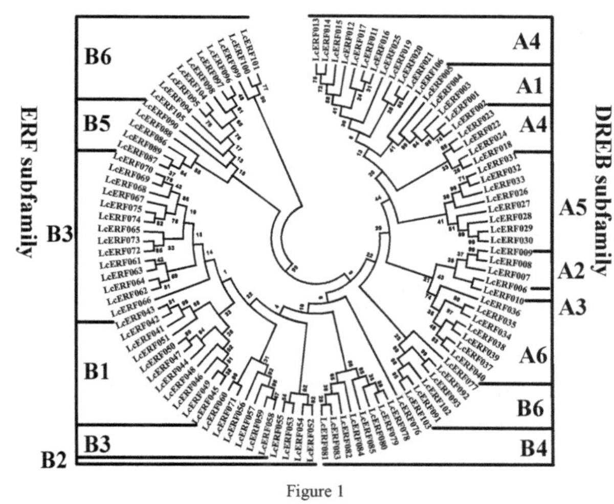

Fig. 6.1 An unrooted phylogenetic tree of *L. japonicus* 106 ERF proteins. The proteins are named according to LcERF numbers

6.6.2 *Screening the Salt Stress-Responsive* LcERF *Genes*

To identify the salt stress responsive *LcERF* genes, 106 *LcERF* genes were investigated using qRT-PCR method. As shown in Fig. 6.2, 45 *LcERF* genes, including 11 *CBF/DREB* genes and 34 *ERF* genes, were responsive to 150 mM NaCl. Whereas among the 45 salt responsive genes, 19 *LcERF* genes increased by 2 ratio, 13 *LcERF* genes had no variation and 13 *LcERF* genes decreased to half after 12 h salt stress compared with the control. Especially *LcERF029*, *LcERF069*, *LcERF073*, and *LcERF076* genes increased by 25 ratio to 60 ratio, *LcERF078* genes increased by 300 ratio. *LcERF054* and *LcERF080* gene were increased by six ratio and four ratio under salt stress, respectively. To understand the roles of *LcERF054* and *LcERF080* genes in plant abiotic stress responses, we further analyzed the function of *LcERF054* or *LcERF080* in *Arabidopsis*.

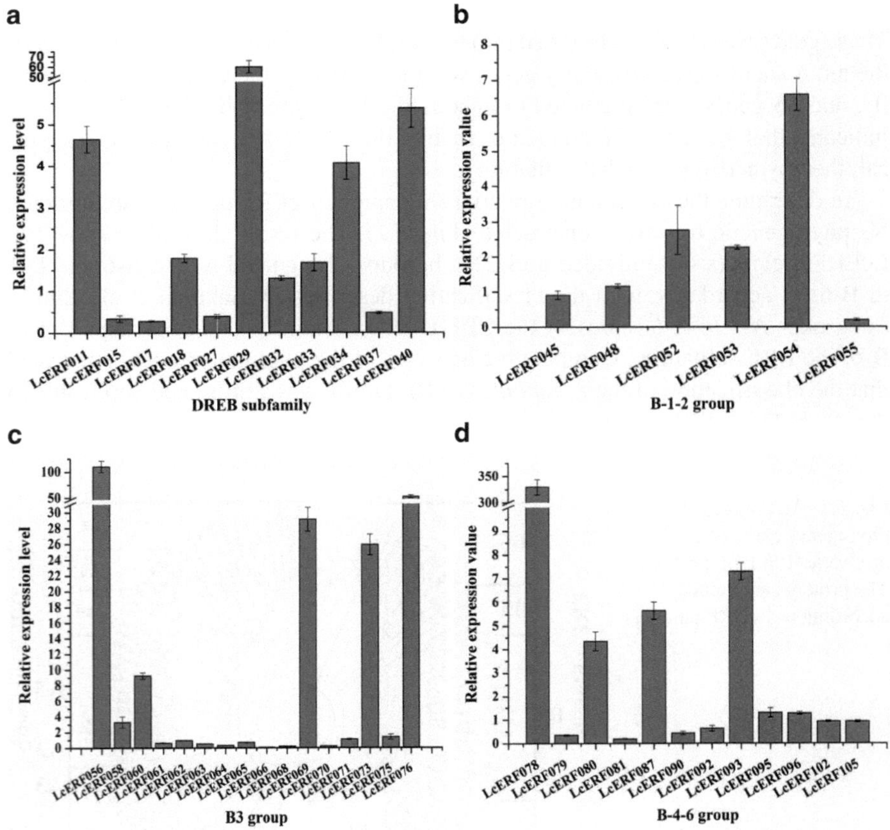

Fig. 6.2 Expression patterns of *L. corniculatus* cv *Leo LcERF* genes under salt treatment. Total RNA was isolated from leaf of *L. corniculatus* cv Leo seedlings with 12 h of salt stress with 150 mM NaCl. In all, 2 μg of total RNA was reverse transcribed into first-strand cDNA for qPCR. The *L. corniculatus* cv Leo ubiquitin gene was used as internal reference

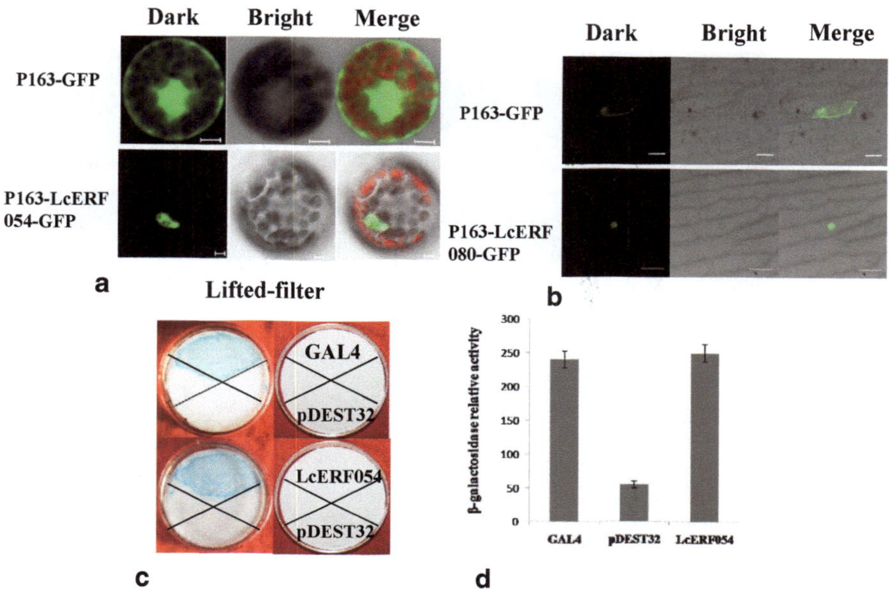

Fig. 6.3 **a** Subcellular localization of LcERF054 protein. Bars = 5 μm. **b** Subcellular localization of LcERF080 protein. Bars = 100 μm. **c** Analysis of transcriptional activation of LcERF054 protein by the yeast one-hybrid assay. **d** The β-galactosidase relative activity assay, Bars indicate standard error of the mean

6.7 LcERF054 and LcERF080 Localizes to the Nucleus and Acts as a Transcriptional Activator

To provide further evidence for the potential role of LcERF054 or LcERF080 in transcriptional regulation, the localization of LcERF054 and LcERF080 protein was determined. LcERF054/LcERF080-GFP fluorescence was observed exclusively in the nucleus, (Fig. 6.3a, b), whereas the control GFP protein was distributed throughout the cell, as expected (Fig. 6.3a, b). These results confirmed that LcERF054 and LcERF080 are nuclear proteins. The transcriptional activation ability of LcERF054 was tested by yeast one-hybrid assay. As shown in Fig. 6.3c, d, Filter papers with yeast transformants harbouring the pDEST32-LcERF054 or GAL4 constructs that had been prewetted with 5-bromo-4-chloro-3-indoyl-β-D-galactopyranoside became blue, whereas, the lifted-filter with pDEST32 was still white. Furthermore, the β-galactosidase relative activity of the pDEST32-LcERF054 or GAL4 was more than four times higher than those of the control vector. These results confirmed the transcriptional activity of the LcERF054 protein.

Fig. 6.4. Effect of *LcERF054/LcERF080* overexpression on salt tolerance in transgenic *Arabidopsis*. Ten-day-old seedlings grown on MS medium plates were planted in pot, filled with a 1:1 mixture of vermiculite and humus, and well watered for 20 d. The seedlings were watered 1/2 MS containing NaCl (100, 200, and 300 mM) at intervals of 4 days for three applications and were monitored during the next weeks. The experiment was repeated three times

6.7.1 Overexpression of LcERF054 *and* LcERF080 *Increased Salt Tolerance in Transgenic* Arabidopsis

To investigate the biological function of *LcERF054* and *LcERF080*, transgenic *Arabidopsis* plants were developed. Three lines of *LcERF054*#21, 26, and 28 and two lines *LcERF080*#13 and 25 were selected for further analysis on the basis of the highest transgene level for further analysis, The transgenic plants showed no morphological difference in the vegetative and floral tissues, as compared to WT plants (Fig. 6.4). The significant phenotypic differences between WT and transgenic plants were observed after 8 days of 300 mM NaCl stress, WT were dead, however, survival rate of transgenic *LcERF054* and *LcERF080* plants were 100 % and 90 %, respectively (Fig. 6.4). Furthermore, the improved salt tolerance of the *LcERF054*

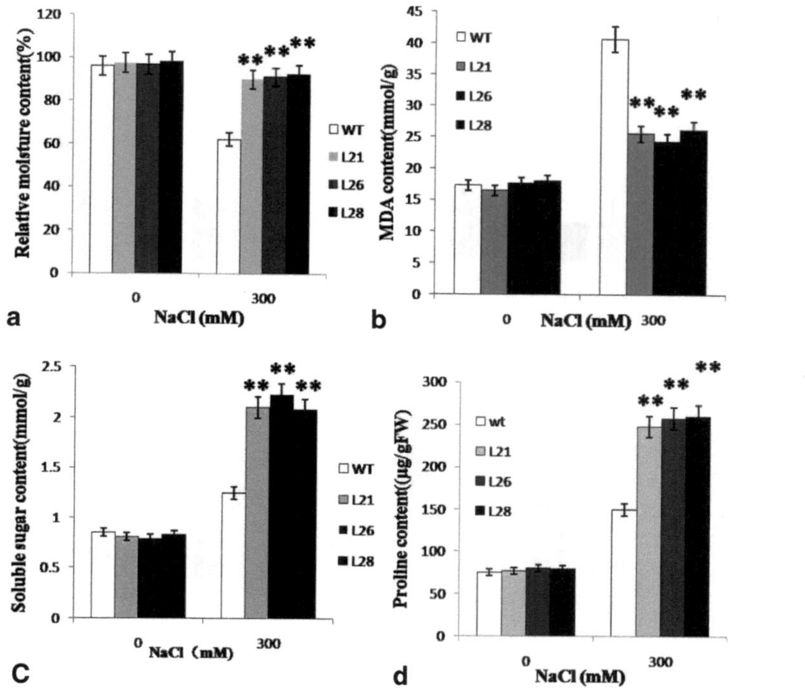

Fig. 6.5. Physiological changes associated with the salt stress response in *LcERF054*-transgenic *Arabidopsis*. **a** Relative moisture. **b** MDA. **c** Soluble sugar. **d** Proline content in seedlings. The data represent the means of three replicates. The error bars indicate the SD; * and ** indicate significant differences in comparison with the WT plants at the $P<0.05$ and $P<0.01$ levels, respectively, using the Student's t-test

and *LcERF080* transgenic plants was correlated with changes in the relative moisture, MDA, soluble sugar and proline contents which are the important parameters for plant responses to abiotic stress (Robles et al. 2012). Our results indicated that no difference in relative moisture, MDA, soluble sugar, and proline content was observed between WT and transgenic plants under standard conditions, however, under high salt stress conditions transgenic plants considerably exhibited higher levels of relative moisture, soluble sugar, and proline content than the WT plants. whereas MDA content was significantly lower in the *LcERF054* and *LcERF080* plants than in WT (Figs. 6.5, 6.6). These results corresponded to the previously described (Zhou et al. 2012; Zhang et al. 2012).

To elucidate the putative molecular mechanisms of *LcERF054* and *LcERF080* in the salt response, we investigated the expression of the seven known salt stress-response gene, including cold-regulated 15A (*COR15A*) (Baker et al. 1994), Δ^1-pyrroline-5-carboxylate synthase (*P5CS1*, *P5CS2*) (Rai and Penna 2013), late embryogenesis abundant (*LEA4-5*) (Hundertmark and Hincha 2008), responsive to dehydration 29A (*RD29A*) (Narusaka et al. 2003) and responsive to dehydration 22 (*RD22*) (Yamaguchi-Shinozaki and Shinozaki 1993) in both transgenic and WT

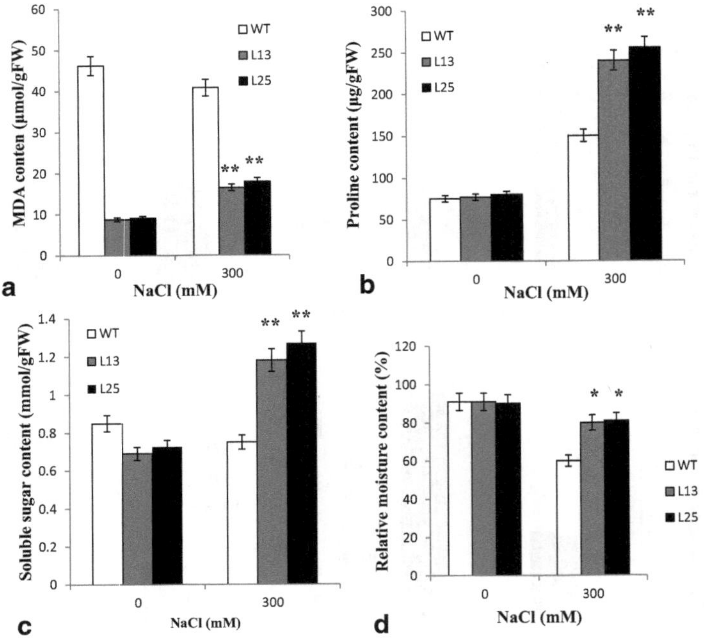

Fig. 6.6 Physiological changes associated with the salt stress response in *LcERF080*-transgenic *Arabidopsis*

plants under normal and salt stress conditions. For *LcERF054*, the transcripts of the target genes was no difference in the transgenic lines compared with the WT under normal condition, excluding *RD29A* (Fig. 6.7a). However, under salt stress condition, four genes, *COR15A*, *P5CS1*, *LEA4–5,* and *RD29A*, exhibited a significantly higher expression level in the *LcERF054* transgenic plants compared to the WT plants (Fig. 6.7b), whereas the expression of the two genes, *RD22* and *P5CS2*, was increased but not significantly. For *LcERF080*, all transcripts were more abundant in *LcERF080*-lines than in WT under standard conditions, excluding *P5CS2* and *RAB18* (Fig. 6.8a). Under salt stress condition, three genes, *COR15A*, *P5CS1*, and *RD22*, exhibited significantly higher expression levels in the *LcERF080* transgenic plants compared to the WT plants (Fig. 6.8b). These data indicated that *LcERF054* and *LcERF080* functioned to activate the expression of these stress-related genes.

6.8 Discussion

6.8.1 Characterization of the AP2/ERF Superfamily from L. japonicus

Recently, the gene structures, phylogeny, and conserved motifs of the AP2/ERF gene family were widely investigated in *Arabidopsis* and rice (Nakano et al. 2006)

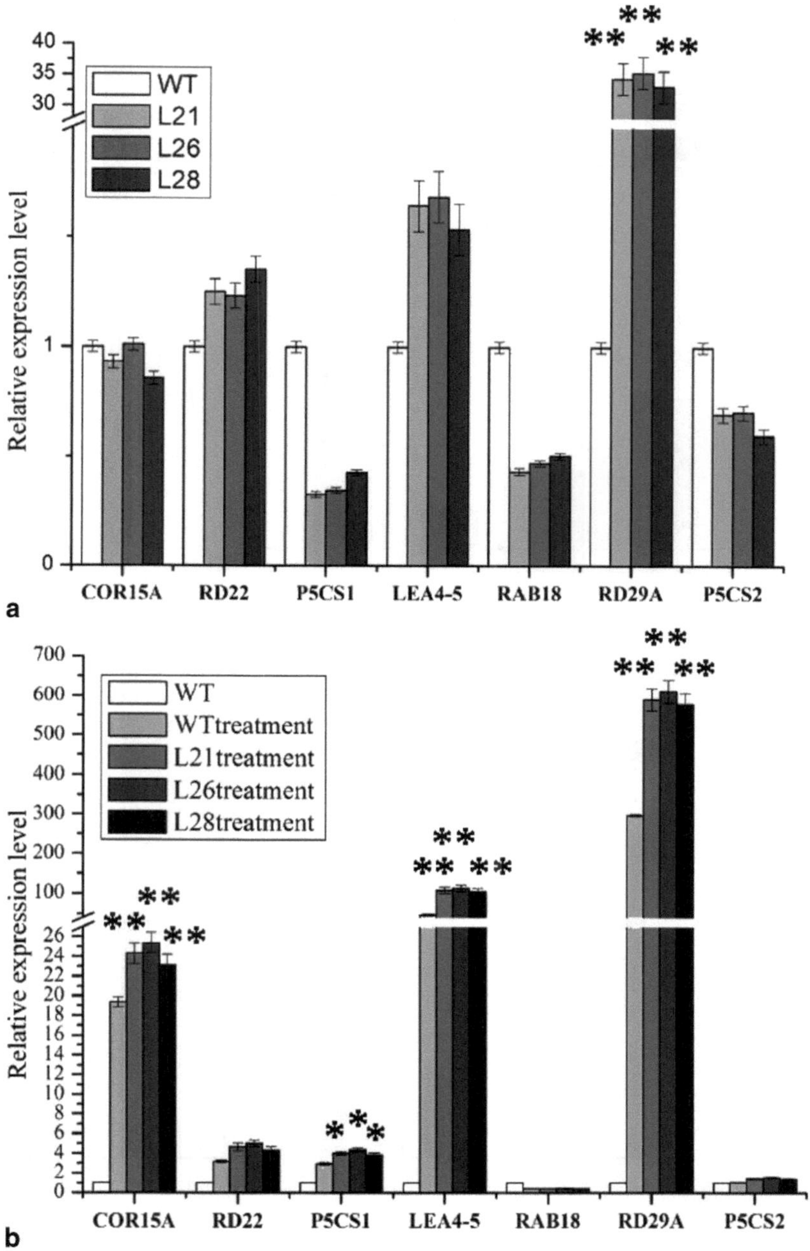

Fig. 6.7 Expression patterns of salt stress-response genes in *LcERF054*-transgenic *Arabidopsis*. Total RNA was isolated from 4-week-old seedlings that were grown under normal and 300 mM salt conditions for 3-day qPCR. Analyses were performed with gene-specific primers. The data represent the means of three replicates. Error bars indicate SD. * and ** indicate significant differences in comparison with the WT plants at $P<0.05$ and $P<0.01$, respectively (Student's t-test)

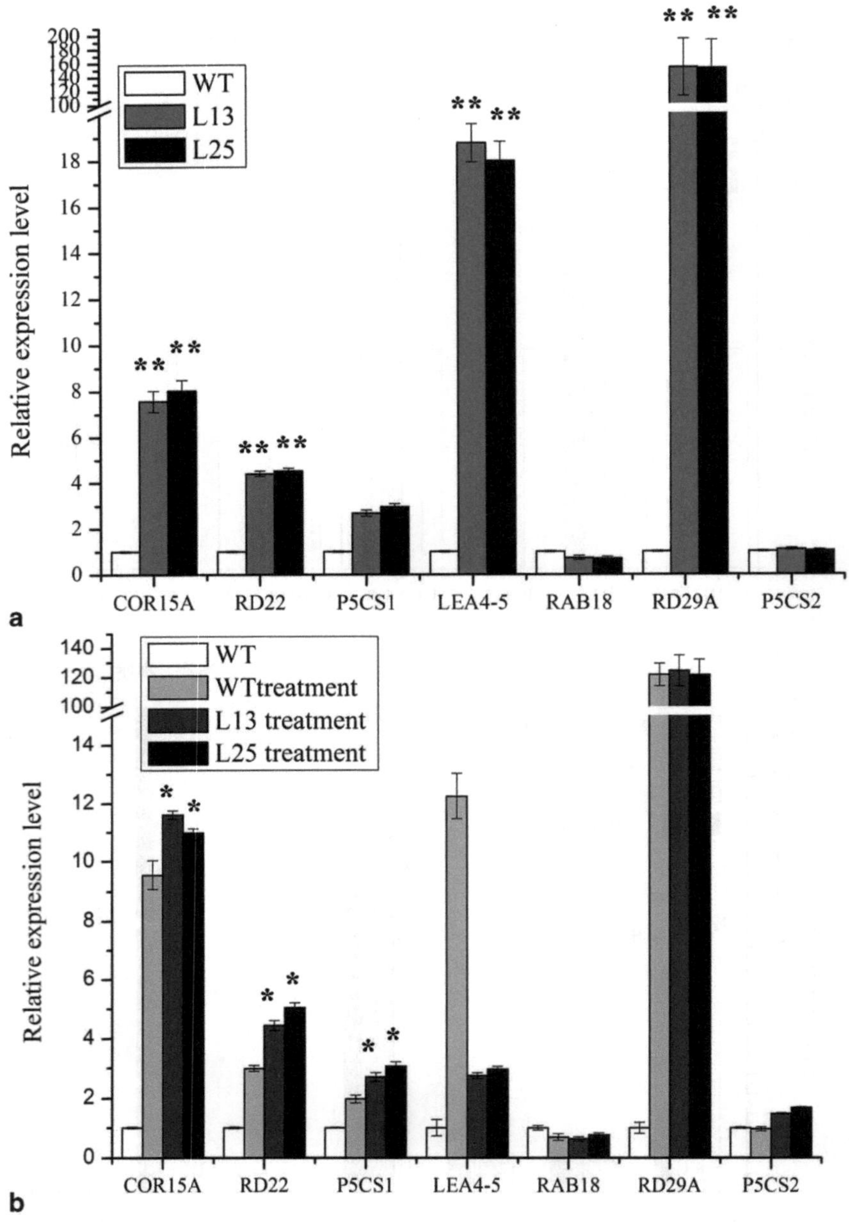

Fig. 6.8 Total RNA was isolated from 4-week-old seedlings that were grown under normal and 300 mM were salt conditions for 3-day qPCR. Analyses were performed with gene-specific primers. The data represent the means of three replicates. Error bars indicate SD. * and ** indicate significant differences in comparison with the WT plants at $P < 0.05$ and $P < 0.01$, respectively (Student's t-test)

and so on. However, there is still little information about *L. japonicus* AP2/ERF genes. To gain further information about the AP2/ERF family in *L. japonicus*, 127 members of the AP2/ERF superfamily were identified from the *L. japonicus* genome DNA database.

Regions outside the DNA-binding domain in transcription factors generally contain either functionally important domains, or motifs associated with transcription regulation and nuclear localization (Kasuga et al. 1999). It has been reported that regions of acidic amino acid-rich, Gln-rich, Pro-rich, Phe-rich, and/or Ser/Thr-rich amino acid sequences, can usually be designated as transcriptional-activation domains. The conserved motifs identified in this study have similar features, such as Ser-rich in LcERF011, LcERF087, and so on. In addition to common conserved motifs in *L. japonicus*, *Arabidopsis*, and rice, there are more *L. japonicus*-specific ERF family motifs, which may have important roles in regulating biological processes, those functions need to be demonstrated further.

6.8.2 Overexpression of LcERF054 *and* LcERF080 *Increased Tolerance to Salt Stress in Transgenic* Arabidopsis *Plants*

Many plant species naturally accumulate proline and soluble sugar as major organic osmolytes when subjected to different abiotic stresses (Ashraf and Foolad 2007). In this study, we found proline and soluble sugar contents increased in the *LcERF054* and *LcERF080* transgenic *Arabidopsis* than in the WT, which suggested that proline and soluble sugars are factors that contribute to the tolerance of *LcERF054* and *LcERF080* transgenic plants to salt stress. It has been widely reported that accumulation of MDA are indicators of free radical production and cellular membrane injury caused by various stresses (Zhou et al. 2012; Robles et al. 2012). In this study, when the plants were grown under salt-induced stress conditions, the MDA content was lower in the *LcERF054* and *LcERF080* transgenic plants than in the WT plants. These results indicate that the LcERF054 or LcERF080 protein is one of the most important salt-induced ERF genes in *L. corniculatus* cv Leo, and plays a crucial role under salt tolerance through changing a variety of physiological and biochemical processes.

Many genes responding to abiotic stress have been used as markers for monitoring stress response pathways in plants, such as *COR15A*, *RD29A*, *P5CS1*, *P5CS2,* and so on. *P5CS1* and *P5CS2* was mainly induced under salt, drought and *P5CS2* was also constitutive housekeeping gene (Rai and Penna 2013), in this study, *P5CS1* transcripts were higher in *LcERF054* and *LcERF080* overexpression lines than that in wild-type under salt stress and normal condition, However, *P5CS2* transcripts showed no difference between *LcERF054* overexpression lines and wild-type under normal/salt stress condition. This result confirmed that *LcERF054* or *LcERF080* overexpression lines increased proline contents by *P5CS1* but not *P5CS2* to enchance salt stress. *RD29A* is induced by high-salinity, desiccation (Narusaka et al. 2003) and exhibited a significantly higher expression level in the *LcERF054* transgenic plants compared to the WT plants under normal and salt stress conditions These results suggested that LcERF054 and LcERF080 could modulate directly or indirectly salt, drought, cold and ABA-induced gene expression such as *COR15A*, *RD29A*, *P5CS1,* and *LEA4–5* to enhance tolerance to salt stress. The mechanism whereby *LcERF054* and *LcERF080* genes confer tolerances when subjected to salt stress and the effect of overexpression in *L. japonicus* needs further investigation.

Acknowledgments This work was supported by the National Natural Science Foundation of China (Grant No. 31372361) and the National Program on Key Basic Research Project (973 Program) (Grant No. 2014CB138701).

References

Agarwal PK, Agarwal P, Reddy MK, Sopory SK (2006) Role of DREB transcription factors in abiotic and biotic stress tolerance in plants. Plant Cell Rep 25(12):1263–1274

Ashraf M, Foolad MR (2007) Roles of glycine betaine and proline in improving plant abiotic stress resistance. Environ Exp Bot 59(2):206–216

Bailey TL, Bodén M, Buske FA, Frith M, Grant CE, Clementi L, et al (2009) MEME suite: tools for motif discovery and searching. Nucleic Acids Res 37(suppl 2):W202–W208

Baker S, Wilhelm K, Thomashow M (1994) The 5′-region of *Arabidopsis thaliana cor15a* has *cis*-acting elements that confer cold-, drought- and ABA-regulated gene expression. Plant Mol Biol 24(5):701–713

Bates LS, Waldren R, Teare ID (1973) Rapid determination of free proline for water-stress studies. Plant Soil 39:205–207

Chen M, Li YP, Chen J, Chen B, Zhou DJ (2003) Construction of yeast expression plasmid hERR1/HBD and preliminary identification of its function. Acta Academiae Medicinae Militaris Tertiae 1:51–53 (in Chinese)

Chen T, Yang Q, Gruber M, Kang J, Sun Y, Ding W, Zhang T, Zhang X (2012) Expression of an alfalfa (*Medicago sativa* L.) ethylene response factor gene *MsERF8* in tobacco plants enhances resistance to salinity. Mol Biol Rep 39(5):6067–6075

Chenna R (2003) Multiple sequence alignment with the Clustal series of programs. Nucleic Acids Res 31(13):3497–3500

Clough SJ, Bent AF (1998). Floral dip: a simplified method for *Agrobacterium*-mediated transformation of *Arabidopsis thaliana*. Plant J 16(6):735–743

Dinh TT, Girke T, Liu X, Yant L, Schmid M, Chen X (2012) The floral homeotic protein APETALA2 recognizes and acts through an AT-rich sequence element. Development 139(11):1978–1986

Duan B, Lu Y, Yin C, Junttila O, Li C (2005) Physiological responses to drought and shade in two contrasting *Picea asperata* populations. Physiol Plantarum 124(4):476–484

Guo A, He K, Liu D, Bai S, Gu X, Wei L, Luo L (2005) DATF: a database of *Arabidopsis* transcription factors. Bioinformatics 21(10):2568–2569

Hao D (1998) Unique mode of GCC box recognition by the DNA-binding domain of Ethylene-responsive Element-binding Factor (ERF Domain) in plant. J Biol Chem 273(41):26857–26861

Huh SU, Lee IJ, Ham BK, Paek KH (2012) *Nicotiana tabacum* Tsip1-interacting ferredoxin 1 affects biotic and abiotic stress resistance. Mol Cells 34(1):43–52

Hundertmark M, Hincha DK (2008) LEA (Late Embryogenesis Abundant) proteins and their encoding genes in *Arabidopsis thaliana*. BMC Genomics 9:118

Jin X, Xue Y, Wang R, Xu R, Bian L, Zhu B, Han H, Peng R, Yao Q (2013) Transcription factor *OsAP21* gene increases salt/drought tolerance in transgenic *Arabidopsis thaliana*. Mol Biol Rep 40(2):1743–1752

Kasuga M, Liu Q, Miura S, Yamaguchi-Shinozaki K, Shinozaki K (1999) Improving plant drought, salt, and freezing tolerance by gene transfer of a single stress-inducible transcription factor. Nat Biotechnol 17(3):287–291

Kinkema M, Fan W, Dong X (2000) Nuclear localization of NPR1 is required for activation of *PR* gene expression. Plant Cell 12(12):2339–2350

Lehner A, Corbineau F, Bailly C (2006) Changes in lipid status and glass properties in cotyledons of developing sunflower seeds. Plant Cell Physiol 47(7):818–828

Letunic I, Doerks T, Bork P (2012) SMART 7: recent updates to the protein domain annotation resource. Nucleic Acids Res 40(D1):D302–D305

Livak KJ, Schmittgen TD (2001) Analysis of relative gene expression data using real-time quantitative PCR and the 2-ΔΔCT method. Methods 25(4):402–408

Mishra AK, Puranik S, Bahadur RP, Prasad M (2012) The DNA-binding activity of an AP2 protein is involved in transcriptional regulation of a stress-responsive gene, *SiWD40*, in foxtail millet. Genomics 100(4):252–263

Mizoi J, Shinozaki K, Yamaguchi-Shinozaki K (2012). AP2/ERF family transcription factors in plant abiotic stress responses. Biochim Biophys Acta 1819(2):86–96

Nakano T, Suzuki K, Fujimura T, Shinshi H (2006) Genome-wide analysis of the ERF gene family in *Arabidopsis* and rice. Plant Physiol 140(2):411–432

Narusaka Y, Nakashima K, Shinwari ZK, Sakuma Y, Furihata T, Abe H, et al (2003) Interaction between two *cis*-acting elements, ABRE and DRE, in ABA-dependent expression of *Arabidopsis rd29A* gene in response to dehydration and high-salinity stresses. Plant J 34(2):137–148

Ohto MA, Floyd SK, Fischer RL, Goldberg RB, Harada JJ (2009) Effects of APETALA2 on embryo, endosperm, and seed coat development determine seed size in *Arabidopsis*. Sex Plant Reprod 22(4):277–289

Pan Y, Seymour GB, Lu C, Hu Z, Chen X, Chen G (2012) An ethylene response factor (ERF5) promoting adaptation to drought and salt tolerance in tomato. Plant Cell Rep 31(2):349–360

Rai AN, Penna S (2013) Molecular evolution of plant *P5CS* gene involved in proline biosynthesis. Mol Biol Rep 40(11):6429–6435

Ripoll JJ, Roeder AH, Ditta GS, Yanofsky MF (2011) A novel role for the floral homeotic gene APETALA2 during *Arabidopsis* fruit development. Development 138(23):5167–5176

Robles P, Micol JL, Quesada V (2012) *Arabidopsis* MDA1, a nuclear-encoded protein, functions in chloroplast development and abiotic stress responses. PLoS One 7(8):e42924

Saitou N, Nei M (1987) The neighbor-joining method: a new method for reconstructing phylogenetic trees. Mol Biol Evol 4(4):405–425

Sakuma Y, Liu Q, Dubouzet JG, Abe H, Shinozaki K, Yamaguchi-Shinozaki K (2002) DNA-binding specificity of the ERF/AP2 domain of *Arabidopsis* DREBs, transcription factors involved in dehydration- and cold-inducible gene expression. Biochem Biophys Res Commun 290(3):998–1009

Sato S, Nakamura Y, Kaneko T, Asamizu E, Kato T, Nakao M, et al (2008) Genome structure of the legume, *Lotus japonicus*. DNA Res 15(4):227–239

Swanson EB, Somers DA, Tomes DT (1990) Birdsfoot Trefoil (*Lotus corniculatus* L.). In: Bajaj YPS (ed) Legumes and oilseed crops I. Springer, Berlin.

Tamura K, Peterson D, Peterson N, Stecher G, Nei M, Kumar S (2011) MEGA5: molecular evolutionary genetics analysis using maximum likelihood, evolutionary distance, and maximum parsimony methods. Mol Biol Evol 28(10):2731–2739

Tian Y, Zhang H, Pan X, Chen X, Zhang Z, Lu X, Huang R (2011) Overexpression of ethylene response factor *TERF2* confers cold tolerance in rice seedlings. Transgenic Res 20(4):857–866

Wurschum T, Gross-Hardt R, Laux T (2006) APETALA2 regulates the stem cell niche in the *Arabidopsis* shoot meristem. Plant Cell 18(2):295–307

Yamaguchi-Shinozaki K, Shinozaki K (1993) The plant hormone abscisic acid mediates the drought-induced expression but not the seed-specific expression of *rd22*, a gene responsive to dehydration stress in *Arabidopsis thaliana*. Molec Gen Genet 238(1–2):17–25

Yoo SD, Cho YH, Sheen J (2007) *Arabidopsis* mesophyll protoplasts: a versatile cell system for transient gene expression analysis. Nat Protoc 2(7):1565–1572

Zhai Y, Wang Y, Li L, Lei T, Yan F, Su L, Li X, Zhao Y, Sun X, Li J, Wang Q (2013) Isolation and molecular characterization of *GmERF7*, a soybean ethylene-response factor that increases salt stress tolerance in tobacco. Gene 513(1):174–183

Zhang L, Zhao G, Xia C, Jia J, Liu X, Kong X (2012) A wheat R2R3-MYB gene, *TaMYB30-B*, improves drought stress tolerance in transgenic *Arabidopsis*. J Exp Bot 63(16):5873–5885

Zhang L, Li Z, Li J, Wang A (2013) Ectopic Overexpression of *SsCBF1*, a CRT/DRE-Binding factor from the nightshade plant *Solanum lycopersicoides*, confers freezing and salt tolerance in transgenic *Arabidopsis*. PLoS One 8(6):e61810

Zhao T, Liang D, Wang P, Liu J, Ma F (2012) Genome-wide analysis and expression profiling of the DREB transcription factor gene family in *Malus* under abiotic stress. Mol Genet Genomics 287(5):423–436

Zhou ML, M JT, Pang JF, Zhang ZL, Tang YX, Wu YM (2010) Regulation of plant stress response by dehydration responsive element binding (DREB) transcription factors. Afr J Biotechnol 9(54):9255–9279

Zhou ML, M JT, Zhao YM, Wei YH, Tang YX, Wu YM (2012) Improvement of drought and salt tolerance in *Arabidopsis* and *Lotus corniculatus* by overexpression of a novel DREB transcription factor from *Populus euphratica*. Gene 506(1):10–17

Chapter 7
Zn-Deficiency Responsive Transcripts in *Agrostis* Species Revealed by mRNA Differential Display

Bala Ani Akpinar, Ozge Canli and Hikmet Budak

7.1 Introduction

Grass family has a particular importance as rice, wheat, and maize constitutes the major portion of diet not only for humans but also for domestic animals. Moreover, grasses form a significant part of the urban and suburban landscape all over the world, covering approximately 20 % of the earth's land surface (Shantz 1954; Kellogg 2001).

Agrostis genus of the Poaceae family consists of more than 200 species. Three bentgrasses, namely Creeping bentgrass (*Agrostis stolonifera*), Colonial bentgrass (*Agrostis capillaris*), and Velvet bentgrass (*Agrostis canina*), are the most widely used members of the genus, especially in athletic fields and golf courses (Hitchcock 1935; Budak et al. 2006). Among these three, creeping bentgrass is the most commercially important bentgrass used especially in golf course due its fine leaves and tolerance to low cutting heights (Belanger et al. 2003).

Zinc (Zn) is one of the eight trace elements which are essential for the normal growth and reproduction of plants. There is a global effort for micronutrient enrichment through plant breeding (Graham et al. 2001). Zn deficiency in soils causes severe reductions in both crop production and grain quality, in countries such as India, Turkey, China, and Australia (Alloway 2004). Studies on various plant species have elaborated many aspects on the importance of metals and the mechanisms of metal accumulation and homeostasis in plants. A recent study on *Medicago truncatula* predicted six new metal binding proteins that have a high similarity to the ZIP protein family (López-Millán et al. 2004). In another study on *Arabidopsis thaliana*, it was reported that cation efflux family is involved in Zn homeostasis (Van der Zaal et al. 1999).

In this study, our goal was to identify differentially expressed gene(s) in response to Zn deficiency in three most widely used bentgrass species; Colonial, Creeping,

H. Budak (✉) · B. A. Akpinar · O. Canli
Molecular Biology, Genetics and Bioengineering, Faculty of Engineering and Natural Sciences,, Sabanci University, 34956 Tuzla, Istanbul, Turkey
e-mail: budak@sabanciuniv.edu

H. Budak, G. Spangenberg (eds.), *Molecular Breeding of Forage and Turf*,
DOI 10.1007/978-3-319-08714-6_7

and Velvet bentgrasses, using mRNA differential display method. Transcripts that exhibited differential expression patterns were sequenced to identify similarities to known proteins, which may account for the Zn deficiency response. One such transcript, exhibiting extensive sequence homology to known plant ZIP Zn transporters, was fully sequenced and its expression levels were quantified under Zn deficiency. This study provides the first Zn responsive transcripts in *Agrostis* species, which may contribute to the exploration and exploitation of Zn-related pathways in detail in this species.

7.2 Materials and Methods

7.2.1 *Plant Material and Growth Conditions*

Three replicates of Creeping bentgrass (*Agrostis stolonifera* L-93 M5-4-872-1), Colonial bentgrass (*Agrostis capillaris* SR7100 B28-6-7100-4), and Velvet bentgrass (*Agrostis canina* 1-464-309) plants were grown under different Zn concentrations in the greenhouse under controlled conditions in 1500 g soil in plastic pots. The Zn deficient soil is obtained from Central Anatolia (pH: 8.04, $CaCO_3$ 14.9 %, organic matter 69 %, salt 0.08 %, and clay 60.6 %). The basal treatment is 200 mg/kg N as $Ca(NO_3)_2$, 100 mg/kg P as KH_2PO_4, 125 mg/kg K as KH_2PO_4, 20 mg/kg S as $CaSO_4.2H_2O$, 2.5 mg/kg F as FeEDTA ($C_{10}H_{12}FeN_2$ NaO_8), and 0, 0.1, and 5 mg/kg Zn as $ZnSO_4.7H_2O$ applied for all species. The plants were grown for 30 days by daily water supply as deionized water. The shoots were harvested and dried at 70 °C for determination of shoot dry matter production and elementary decomposition of the shoot. For the molecular analysis, the shoots of the plants grown in 0 and 5 mg/kg Zn were frozen in liquid nitrogen and kept at − 80 °C until use.

The dried shoot samples were grounded and 0.05–0.2 g of grounded samples was ashed at 500 °C for 12 h for determination of elementary concentration. The ashed samples were dissolved in 3.3 % HNO_3 (v/v). The concentration of Zn as well as other elements (Ca, K, P, S, Mg, Fe, Mn, Cu, Al, and Na) were measured by inductively coupled plasma optical emission spectroscopy (ICP-OES, Varian, Australia) at 214.439 nm emission wavelength.

For tissue culture experiments of Creeping bentgrass, ten seeds, after surface sterilization with 5 % sodium hypochlorite for 10 min, were planted in 20 ml plant tissue culture medium containing 0 or 5 ppm Zn. Plants were grown for 21 days in the growth chamber at 26 °C and with 16:8 h of light and dark photoperiod.

7.2.2 *Application of mRNADD and Sequence Analysis*

500 ng of cDNAs obtained by Omniscript Reverse Transcription Kit was subjected to mRNA differential polymerase chain reaction (PCR) by 72 combinations of forward (P) and reverse (T) primers (Table 7.1; Liang and Pardee 1992).

Table 7.1 Primers used in mRNA differential display

Primer designation	Sequence (5′ -3′)
P1	ATT AAC CCT CAC TAA ATG CTG GGG A
P2	ATT AAC CCT CAC TAA ATC GGT CAT AG
P3	ATT AAC CCT CAC TAA ATG CTG GTG G
P4	ATT AAC CCT CAC TAA ATG CTG GTA G
P5	ATT AAC CCT CAC TAA ATG CTG GTA G
P6	ATT AAC CCT CAC TAA ATG CTG GGT G
P7	ATT AAC CCT CAC TAA ATG CTG TAT G
P9	ATT AAC CCT CAC TAA ATG TGG CAG G
T1	CAT TAT GCT GAG TGA TAT CTT TTT TTT TAA
T2	CAT TAT GCT GAG TGA TAT CTT TTT TTT TAC
T3	CAT TAT GCT GAG TGA TAT CTT TTT TTT TAG
T4	CAT TAT GCT GAG TGA TAT CTT TTT TTT TCA
T5	CAT TAT GCT GAG TGA TAT CTT TTT TTT TCA
T6	CAT TAT GCT GAG TGA TAT CTT TTT TTT TCG
T7	CAT TAT GCT GAG TGA TAT CTT TTT TTT TGA
T8	CAT TAT GCT GAG TGA TAT CTT TTT TTT TGC
T9	CAT TAT GCT GAG TGA TAT CTT TTT TTT TGG

The differentially amplified PCR fragments were extracted, ligated into PGEM-T easy vector (Promega, Madison, Wisconsin, USA) and sequenced. The sequences obtained were first subjected to VecScreen (NCBI) algorithm in order to eliminate the vector sequences and were then compared with nucleotide and protein sequence databases using BLASTN and BLASTX algorithms (NCBI). Using ORF finder tool of NCBI, open reading frames of the sequences were detected and the amino acid sequence of the corresponding protein was obtained. The protein sequences were searched for known protein motifs using Motif Scan algorithm of SIB—Swiss Institute of Bioinformatics by searching against PROSITE patterns, PROSITE patterns (frequent match producers), PROSITE profiles, Prefile (more profiles), Pfam HMMs (local models), Pfam HMMs (global models) databases (http://myhits.isb-sib.ch/cgi-bin/motif_scan). For pairwise and multiple alignments, CLUSTALW and CLUSTALX (EBI) algorithms were used.

7.2.3 Real Time PCR and RACE Analysis

For real-time PCR analysis, 18S ribosomal RNA was used as a housekeeping gene (Table 7.2). SYBR Green was used for the fluorescence and the amplifications were performed in Biorad iCyclerQ. The GeneRacer Kit (Invitrogen, L1500) was used to obtain 5' and 3' ends of the mRNA fragment from 1 μg of total RNA. The primers used for RACE are given in Table 7.3.

Table 7.2 Primers used for quantitative Real-Time PCR

18Sribosomal RNA Forward	Sequence (5′ -3′)	Amplicon size
18Sribosomal RNA Reverse	ATG ATA ACT CGA CGG ATC GC	200bp
B1 Forward	CTT GGA TGT GGT AGC CGT TT	
B1 Forward	GAC ACC GCC TGA AGC ATC G	309bp
B1 Reverse	GAC ACC GCC TGA AGC ATC G	

Table 7.3 Primers used for RACE

Primer designation	Sequence (5′ -3′)	TM(°C)
GeneRacer5′	CGACTGGAGCACGAGGACACTGA	74
GeneRacer5′nested	GGACACTGACATGGACTGAAGGAGTA	74
GeneRacer3′	GCTGTCAACGATACGCTACGTAACG	74
GeneRacer3′nested	CGCTACGTAACGGCATGACAGTG	72
Genespecific5′	ATTCGCACCATCGCTGTCAA	65
Genespecific5′nested	ATTCGCACCATCGCTGTCAA	65
Genespecific3′	CGACTCGGAAAGCTGCTTGA	66
Genespecific3′nested	CATGGTGTCGGTGGTGCTTG	66

7.3 Results

7.3.1 Physiological Analysis of Agrostis Species Under Zn Deficiency

The concentration and content of Zn in the seeds of the three bentgrasses were measured prior to the greenhouse trial (Table 7.4). The measurements indicated that Colonial and Creeping bentgrass had similar Zn contents per seed, whereas, Velvet bentgrass had a higher Zn content per seed. Due to its higher amount of stored Zn in the seed, Velvet bentgrass was expected to have a higher tolerance to Zn deficiency than the other two genotypes.

The total dry weight of plants per pot for three replicates were measured for Creeping, Colonial, and Velvet bentgrasses under varying Zn concentrations. As shown in Fig. 7.1, dry weight per plant tended to increase in with increasing Zn supply for all species. While the Creeping bentgrass was the most responsive to Zn sup-

Table 7.4 Zinc content and concentrations of the seeds of the three bentgrasses

Species	Seed Zn content (μg/seed)	Seed Zn concentration (mg/kg)
Colonial bentgrass	2.74	47
Creeping bentgrass	2.78	36
Velvet bentgrass	3.85	6

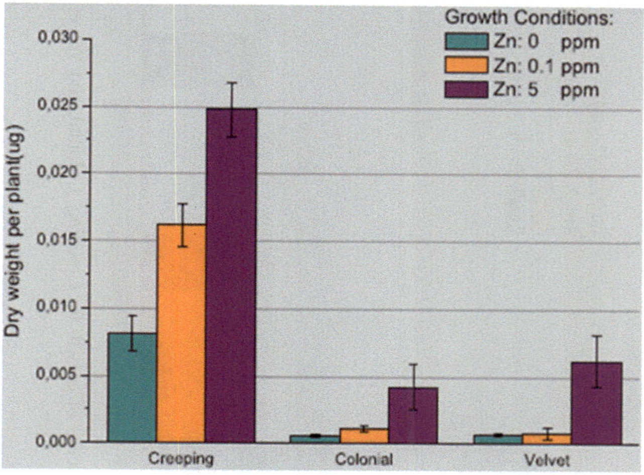

Fig. 7.1 Dry weight measurements per plant in response to different Zn supplies after 4 weeks of growth for Creeping, Colonial and Velvet bentgrasses, obtained from three replicates

ply in terms of dry weight, Colonial and Velvet bentgrasses exhibited considerable increase in dry weight at 5 ppm Zn. The growth of Creeping bentgrass decreased by 67 % when there was no Zn supply and by 35 % when 0.1 ppm Zn was supplied, compared to 5 ppm of Zn supply. On the other hand the growth of Colonial bentgrass decreased by 87 % and by 74 % when supplied with no or 0.1 ppm Zn, respectively, compared to 5 ppm of Zn supply. Similarly, the growth of Velvet bentgrass decreased by 88 % and by 87 % when no or 0.1 ppm Zn was supplied, compared to 5 ppm of Zn supply. The differences in growth of plants under varying Zn supplies suggested that the most tolerant species to Zn deficiency in terms of dry weight was Creeping bentgrass as indicated by higher dry weights under all conditions of Zn supply compared to other species. The growth of Velvet bentgrass was affected slightly more than Colonial bentgrass, however, it was quite unresponsive to 0.1 ppm Zn application whereas Colonial bentgrass was responsive. It can be concluded that, in terms of dry weight, the most Zn deficiency tolerant species was Creeping bentgrass and the most susceptible one was Velvet bentgrass.

The Zn concentration and content of Creeping, Colonial, and Velvet bentgrass after 4 weeks of growth under three different Zn concentrations were measured to gain a better understanding of the Zn supply response (Fig. 7.2). These measurements revealed a similar trend; increasing Zn concentration and content with increasing Zn supply. While the Zn concentration patterns were similar across species under same conditions (Fig. 7.2a), Zn content per plant followed a more similar pattern as the dry weight, where Colonial and Velvet bentgrasses were fairly unresponsive to low Zn supplies (Fig. 7.2b). Although the total Zn content and dry weight of the plants were remarkably higher for Creeping bentgrass, the concentration of Zn in shoots were similar in all three species; in fact, the Zn concentration is lowest in Creeping bentgrass at 5 ppm Zn supply. The unexpected amount of Zn concentration in Colonial bentgrass grown without Zn was probably due to soil contamination during harvesting, as the same trend was not observed in the overall Zn content.

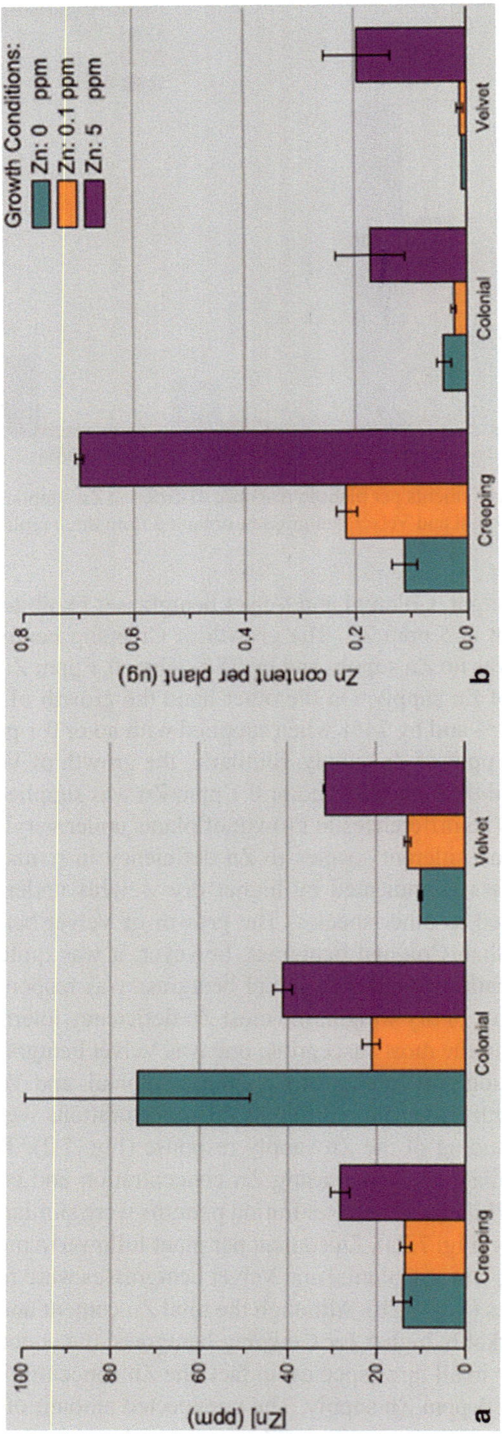

Fig. 7.2 Zn concentration (**a**), and, Zn content (**b**) of plants in response to different Zn supplies after 4 weeks of growth for Creeping, Colonial, and Velvet bentgrasses, obtained from three replicates

7.3.2 Differentially Expressed Genes in Creeping Bentgrass

Through mRNA differential display (mRNADD), six differentially expressed transcripts were identified from Creeping bentgrass at no and 2 ppm Zn supply under sterile tissue culture (Fig. 7.3a–d) and greenhouse conditions (Fig. 7.3e–f). These differentially expressed transcripts amplified with P1T7, P3T1, P4T2, P6T5, P3T2, and P6T3 primer pairs, were extracted and sequenced, corresponding to 156, 415, 199, 175, 145, and 221 bp fragments, respectively. The sequences of the fragments are presented in Table 7.5. The partial mRNA fragment sequences exhibited high similarities with the metal-related proteins in GenPept protein database (Table 7.6).

7.3.3 Identification of ZIP Family Zn Transporter in Creeping Bentgrass

Intriguingly, the fragment amplified with P3-T1 primers in Creeping bentgrass had very significant similarity with the known ZIP Zn transporters. Thus, the differential expression of this transcript in response to Zn deficiency was quantified by quantitative real-time-PCR (qRT-PCR). When the data was normalized in reference

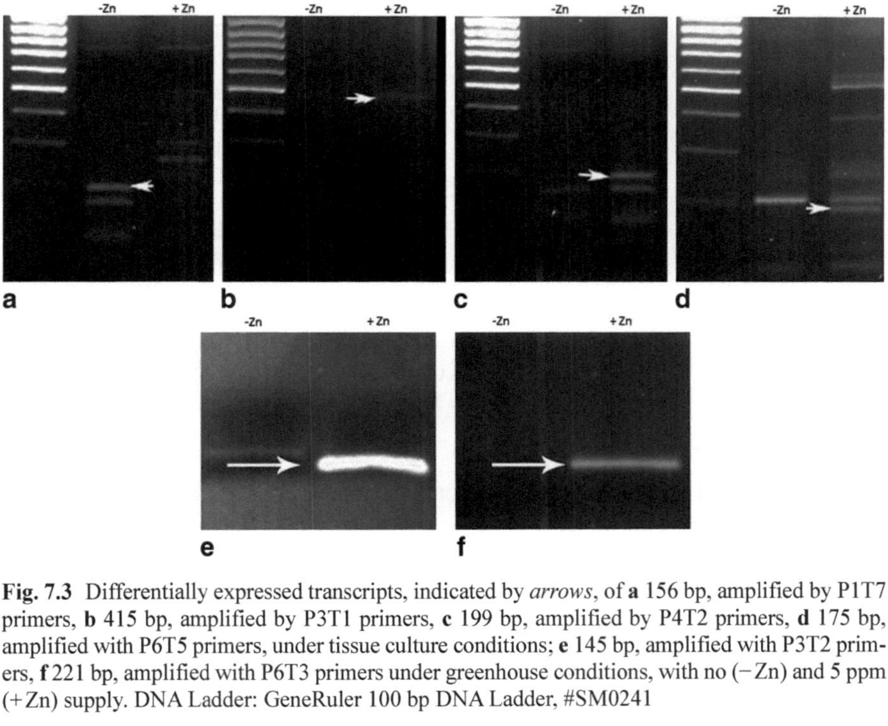

Fig. 7.3 Differentially expressed transcripts, indicated by *arrows*, of **a** 156 bp, amplified by P1T7 primers, **b** 415 bp, amplified by P3T1 primers, **c** 199 bp, amplified by P4T2 primers, **d** 175 bp, amplified with P6T5 primers, under tissue culture conditions; **e** 145 bp, amplified with P3T2 primers, **f** 221 bp, amplified with P6T3 primers under greenhouse conditions, with no (−Zn) and 5 ppm (+Zn) supply. DNA Ladder: GeneRuler 100 bp DNA Ladder, #SM0241

Table 7.5 Sequences of the isolated six differentially expressed fragments from Creeping bent-grass under tissue culture and greenhouse conditions.

Primers	Size(bp)	Sequence
P1-T7	156	GCT CTG AGT AAG AGA CTG AGA TAA AGA TAT CAA AAG ATT CAA GCA TCA CAG TTA TAC CAA AAT TAA TCT TAT TCC AGG ACA AGC ATG TTA CAC TAA GAA TGA CAG CGG TGC TTT CAA GAG GGT GCT CAA AAA AAA AGA TAT CAC TCA GCA TAA TGA
P3-T1	415	AAA GAC CGC CGC CTC CCC GGC GCA CGA CAT GGA CAT TCT GAT CTA TGG TCG GTG GCT CAC GAG GAG GGT GCA CAA CAC ATA CAC ATG GAT GCG CAC ACA ACT GGA GGA CAA CTA TAC TCA CAT GGT GCT ATA CTG GTC TGT GCA GGC TCA CAG TTG GAG GCA GTA CCT ACA ATC AAT TCG CAT ACG TCA GTA CTA TCG GTT CTT ACG CTG TTG GAA TCG GTG CAT TCA GTT ATC ATT GGA GTA TCC TTA AGT GCA TCT GTA CGG CCA TCT TAA GCC AAT CAG GTG CCC TAC TTG CTC AAC CAG CTT CCG TCA GGT CTT TAT GTG TGG ACG TTT GGT TTG CAT TGT CAG TTA CGC TAC TTG GAA CCT ACG GGC GTC ATT GTT TCC TTA AAA AAA AAG ATA TCA CTC AGC ATA ATG A
P4-T2	199	AAT AAG GAT AGA GAG GTT GCT GTT AGA TGT AGT AAT AGA AGA CAT ATT TAT CAA TGT ACT GAA ACA AAA AAA TCT AAA TAT AAT ACC CTT GAG GGG TTC CTC CTT GAA GTG TTG CAT TGT TAT TTA ATG GTT TCT AGG AGT ATG GAG AAA ATG TTT GCC CTC TAA TTG GTA AAA AAA AAG ATA TCA CTC AGC ATA ATG A
P6-T5	175	GGC GAG GGA AGA ACT GCT GAA GAT ATA AGT ANA AGC CCA CCC CAA GAT GAG TGC TTT CTC CTC CGA CTT CCC TAG AGC CTA CGG TGT CAC AGC CGA AAC AGC GAC GGG TTC TCC ACC CAT ATG GGG ATG GAG CGA CAG AAG TAT GGA AAA AAA AAG ATA TCA CTC AGC ATA ATG A
P3-T2	145	TCA TTA CTG CTG AGT GAT ATC TTT TTT TTT ACA CAT ACT GAT CAC AGG TCC CAC CTA TGA AGT TGA TCC GGG GCA CAA CGC TCA GTG TGA TCA GAC GGG ATT CGA CTA TCA TGG ACT TCC CAC CAG CAT TTA GTG AGG GTT AAT A
P6-T3	221	TAT TCG ATA TCA CTA GTG AAT TCG CGG CCC GCC TGC AGG TCG ACC ATA CTG GGA GAG CTC CCA ACG CGT TGG ATG CAT AGC TTG AGT ATT CTA TAG TGT CAC CTA AAT AGC TTG AGC GTA ATC ATG GTC ATA GCT GTT CCT GTG TGA AAT TGT ATC CGC TCA CAA TTC CAC ACA ACA TAC GAG CCG GAA GCA TAA AGT GTA AAG CCT GGG GTG CCT AAT GA

to 18S ribosomal RNA expression level, approximately 25- and 26-fold downregulation of the expression level was observed under Zn deficiency in sterile tissue culture and greenhouse conditions, respectively (Figs. 7.4, 7.5).

The full length mRNA of the fragment was obtained using 5' and 3' rapid amplification of cDNA ends (RACE) (Table 7.7). The open reading frame (ORF) region in the +1 frame, of 238 amino acid long, was found to have 70% identity with Zn transporter ZIP3 identified in *Oryza sativa* (gb: AAP85537.1) in

Table 7.6 BlastX results of the isolated six differentially expressed fragments under tissue culture and greenhouse conditions using NCBI BLASTX algorithm with Blossom80 and PAM30 matrices

Primers	BLASTX hit	Identity	Frame
P1-T7	gi:110598354 conserved hypothetical protein [*Chlorobium ferrooxidans* DSM 13031]	44 %	+3
P3-T1	gb:AAP85537.1 zinc transporter ZIP3 [*Oryza sativa* (*japonica* cultivar-group)]	52 %	+1
	gb:EAY95519.1 hypothetical protein OsI016752 [*Oryza sativa* (*indica* cultivar-group)]	47 %	+1
	gb:AAW68439.1 zinc transporter ZIP [*Triticum aestivum*]	47 %	+1
P4-T2	gb:AAD20819.1 putative casein kinase I [*Dendrobium grex* Madame Thong-In]	100 %	−3
P6-T5	gb:AAU43927.1 putative polyprotein [*Oryza sativa* (*japonica* cultivar-group)]	43 %	+2
	gb:ABF13594.1 HtaA-like surface protein [*Propionibacterium acnes*]	40 %	+3
	ref:XP001390958.1 hypothetical protein An06g01360 [*Aspergillus niger*]	46 %	+3
	ref:YP305367.1 F420-nonreducing hydrogenase II, large subunit [*Methanosarcina barkeri* str. fusaro]	48 %	+3
	ref:NP611847.2 CG5591-PA [*Drosophila melanogaster*]	50 %	−1
	ref:YP983193.1 Ferrochelatase [*Polaromonas naphthalenivorans* CJ2]	57 %	−2
P3-T2	gb:EAZ45719.1 hypothetical protein OsJ029202 [*Oryza sativa* (*japonica* cultivar-group)]	83 %	+2
	dbj:BAD46211.1 unknown protein [*Oryza sativa* (*japonica* cultivar-group)]	83 %	+2
	ref:ZP00518540.1 CBS:Peptidase M50 [*Crocosphaera watsonii* WH 8501]	90 %	+1
	ref:YP001088149.1 putative iron compound ABC transporter, permease protein [*Clostridium dicile* 630]	84 %	−1
	ref:ZP01803509.1 hypothetical protein CdifQ04001892 [*Clostridium dicile* QCD-32g58]	84 %	−1
	ref:YP206783.1 ferric anguibactin transport system permease protein FatC [*Vibrio scheri* ES114]	84 %	−1
P6-T3	emb:CAD42938.2 manganese superoxide dismutase [*Taiwanofungus camphoratus*]	90 %	−3
	ref:YP775527.1 putative transcriptional regulator, Crp/Fnr family [*Burkholderia cepacia* AMMD]	78 %	−2

BLASTX results and showed high similarities in the pairwise alignment (Fig. 7.6a). The sequence was highly conserved in the conserved regions of plant ZIP family Zn transporters (Fig. 7.6b). The cladogram tree of the multiple alignment indicated that the protein was mostly similar to ZIP family Zn transporters of *Oryza sativa* (GI47169687) and *Triticum aestivum* (GI58221593) (Fig. 7.6c).

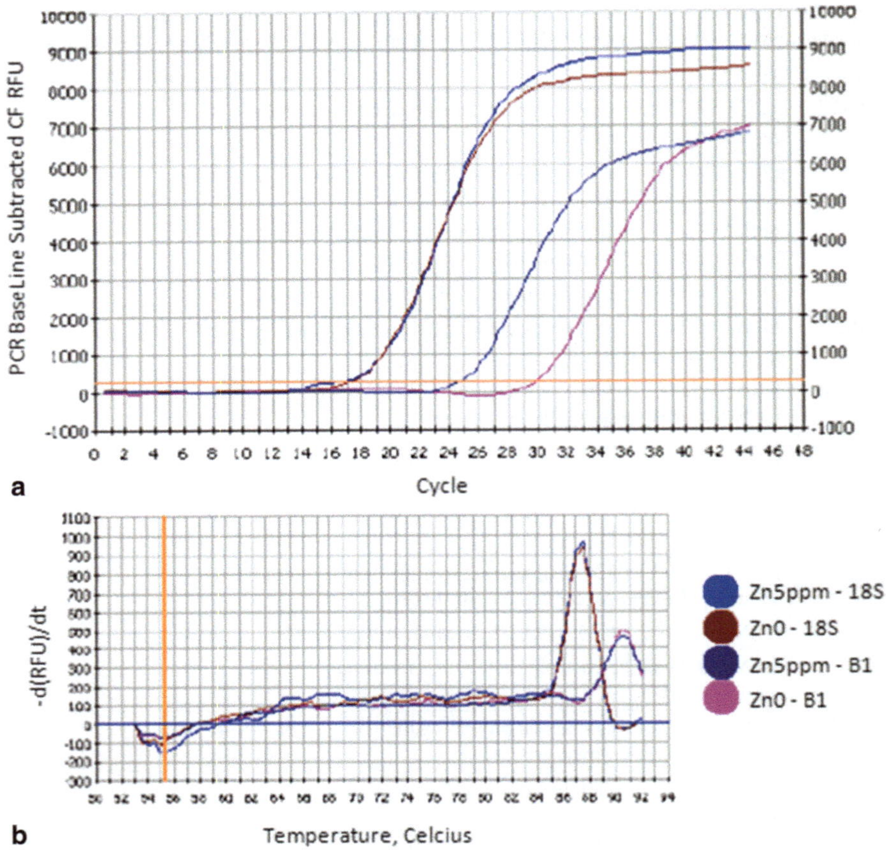

Fig. 7.4 Quantitative real-time amplification plots of the fragment amplified with P3-T1 primers in Creeping bentgrass grown in tissue culture. 18S ribosomal RNA was used as a reference gene. **a** The PCR Base Line Subtracted Curve Fit Data with calculated threshold using the maximum curvature approach is 266.9, per-well baseline cycles have been determined automatically, data analysis window is set at 95 % of a cycle, centered at end of the cycle, weighted mean digital filtering being applied and global filtering is on, **b** The melt curve with weighted mean digital filtering applied, global fltering is on and threshold for automatic peak detection is set at 1.00

7.3.4 Colonial and Velvet Bentgrass mRNADD Identified Additional Differentially Expressed Transcripts

The mRNADD identified a 128 bp fragment whose expression level increased in response to Zn deficiency in Colonial bentgrass (Fig. 7.7a). This fragment was extracted and sequenced, which revealed high similarities to a protein described as elongation of very long chain fatty acids (FEN1/Elo2, SUR4/Elo3,yeast)-like 4 isolated from *Homo sapiens*, a region of putative myosin heavy chain identified in *Dendrobium grex* Madame Thong-In and CRK6 (CYSTEINE-RICH RLK 6) of *Arabidopsis thaliana* that is known as a member of phosphotransferases of the serine or threonine-specific kinase subfamily (Table 7.8).

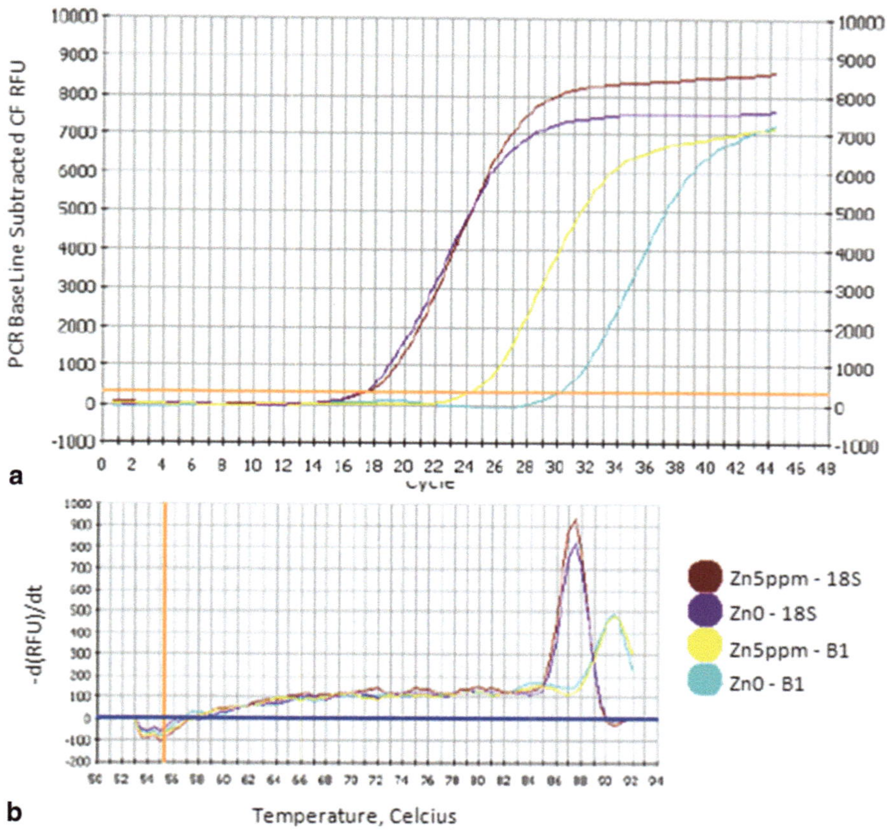

Fig. 7.5 Quantitative Real-Time amplification plots of the fragment amplified with P3-T1 primers in Creeping bentgrass grown in greenhouse. 18S ribosomal RNA was used as a reference gene. **a** The PCR Base Line Subtracted Curve Fit Data with calculated threshold using the maximum curvature approach is 266.9, per-well baseline cycles have been determined automatically, data analysis window is set at 95 % of a cycle, centered at end of the cycle, weighted mean digital filtering being applied and global filtering is on, **b** The melt curve with weighted mean digital filtering applied, global filtering is on and threshold for automatic peak detection is set at 1.00

In Velvet bentgrass, two differentially expressed genes were detected in response to Zn deficiency via mRNADD. The fragments amplified with P2T5 and P3T8 primers exhibited upregulation and downregulation, respectively, Zn deficiency (Fig. 7.7b). The fragment amplified with P2T5 primers revealed 60 % identity to cytochrome c oxidase subunit III of *Aleurodicus dugesii* and 100 % identity to a region of putative casein kinase I of *Dendrobium grex* Madame Thong-In (Table 7.8). The fragment amplified with P3T8 primers showed 61 % identity to putative copper/Zn superoxide dismutase copper chaperone of *Dendrobium grex* Madame Thong-In. In addition, the same fragment exhibited 71 % identity to glutathione-independent formaldehyde dehydrogenase of *Methanosarcina mazei*, which has a Zn-binding dehydrogenase domain (Table 7.8). Differentially expressed transcript sequences are given in Table 7.9.

Table 7.7 The full mRNA sequence of putative ZIP zinc transporter identified in Creeping bentgrass obtained by RACE

Primers	Size(bp)	Sequence
P3-T1	1009	CCG CGC GAC CGT CAG CGA ATT ACG GGC AAT CAA TCC ACA TAC TCA ACA ATT ATG GGA GCC AAG AAG CAT ACC TTG CAA GTG CTT CCA TGG CTA CTG CTC TTT GCG CAG CAC ACT GCA GCC AGT GCC TGC GAC TGT GCT AAC ACC ACA GAC GGG GCA GAC AGA CAG GGT GCA ATG AAG CTA AAG CTC ATT GCC ATT GCA TCC ATC CTT GCA GCT GGG GCG GCT GGT GTC TTG GTG CCA GTG ATT GGC CGC TCC ATG GCT GCG CTG CGC CCT GAT GGT GAC ATC TTC TTT GCT GTC AAG GCA TTT GCA GCC GGC GTC ATC CTT GCC ACT GGC ATG GTG CAC ATT CTT CCA GCG GCG TTT GAT GCG CTC ACA TCT CCA TGC CTC AAA AGG GGT GGT GGG GAT AGG AAT CCC TTC CCC TTT GCG GGC CTC AGA CTC GCT GGC TGC TGG GTA CTA CCG CCG GTC TCA CTT GTG GTG GGG ATA GGA ATC CCT TCC CCT TTG AAA GAC CGC CGC CTC CCC GGC GCA CGA CAT GGA CAT TCT GAT CTA TGG TCG GTG GCT CAC GAG GAG GGT GCA CAA CAC ATA CAC ATG GAT GCG CAC ACA ACT GGA GGA CAA CTA TAC TCA CAT GGT GCT ATA CTG GTC TGT GCA GGC TCA CAG TTG GAG GCA GTA CCT ACA ATC AAT TCG CAT ACG TCA GTA CTA TCG GTT CTT ACG CTG TTG GAA TCG GTG CAT TCA GTT ATC ATT GGA GTA TCC TTA AGT GCA TCT GTA CGG CCA TCT TAA GCC AAT CAG GTG CCC TAC TTG CTC AAC CAG CTT CCG TCA GGT CTT TAT GTG TGG ACG TTT GGT TTG CAT TGT CAG TTA CGC TAC TTG GAA CCT ACG GGC GTC ATT GTT TCC TTA AAA AAA AAG ATA TCA CTC AGC ATA ATG ACT AGA TTT TCT AAA ACC AAA GCT TAT CTC AGG TGA TCT CTT TGT ACT TTA TGA TAT GAT AAT ACA TGA TGA CTT AAC TTG TTC TTA AAA AAA AAA AAA A

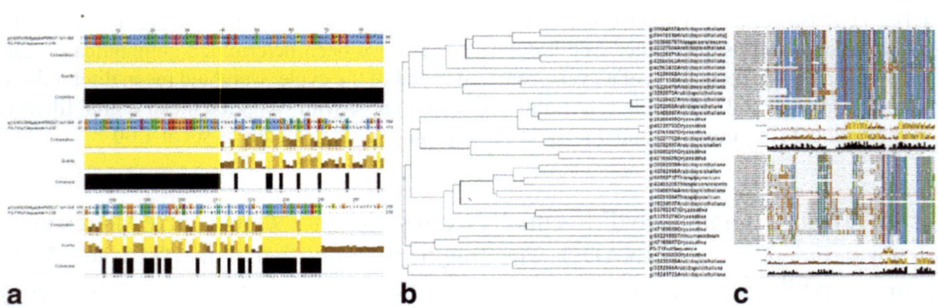

Fig. 7.6 Sequence similarities of full-length transcript identified by RACE in Creeping bentgrass. **a** The pairwise alignment results of the protein product of the full mRNA sequence and *Oryza sativa* zinc transporter ZIP3 (gb:AAP85537.1) using ClustalW algorithm, **b** The multiple alignment of the protein product of the full mRNA sequence and the known ZIP family proteins of various plants using ClustalX algorithm, **c** The cladogram tree of the ClustalX multiple alignments of the protein product of the full mRNA sequence and the known ZIP family proteins of plants

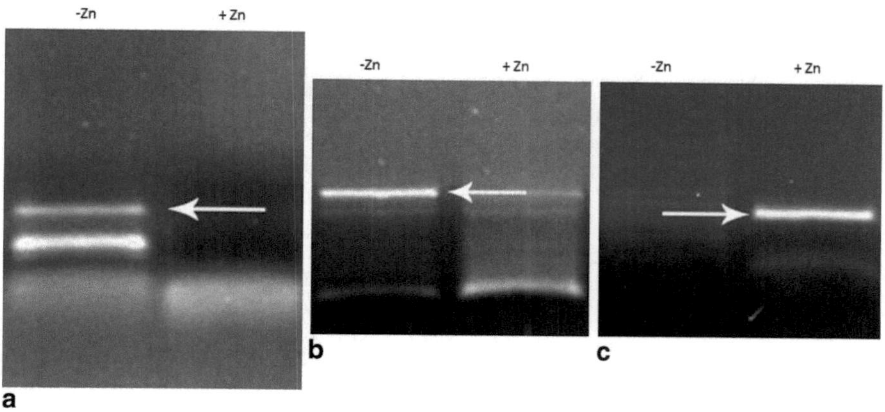

Fig. 7.7 Differentially expressed transcripts, indicated by *arrows*, isolated from Colonial bentgrass **(a)** 128 bp, amplified by P3T6 primers, and isolated from Velvet bentgrass **(b)** 327 bp, amplified by P2T5 primers, **(c)** 146 bp, amplified by P3T8 primers, under greenhouse conditions, with no (-Zn) and 5 ppm (+Zn) supply

Table 7.8 BlastX results of sequenced fragments isolated from Colonial and Velvet bentgrasses using NCBI BLASTX algorithm with Blossom80 and PAM30 matrices

Primers		BLASTX hit	Identity	Frame
P2-T5	Colonial bentgrass	gb:AAH38506.1 Elongation of very long chain fatty acids (FEN1/Elo2, SUR4/Elo3,yeast)-like 4 [*Homo sapiens*]	85 %	+2
		gb:AAD20814.1 putative myosin heavy chain [*Dendrobium grex* Madame Thong-In]	100 %	+3
		ref:NP849426.1 CRK6 (CYSTEINE-RICH RLK 6) [*Arabidopsis thaliana*]	87 %	+3
P2-T5	Velvet bentgrass	ref:YP026057.1 cytochrome c oxidase subunit III [*Aleurodicus dugesii*]	60 %	+3
		gb:AAD20819.1 putative casein kinase I [*Dendrobium grex* Madame Thong-In]	100 %	−3
P3-T8	Velvet bentgrass	gb:AAC79870.1 putative copper/zinc superoxide dismutase copper chaperone [*Dendrobium grex* Madame Thong-In]	61 %	+1
		ref:NP632773.1 Glutathione-independent formaldehyde dehydrogenase [*Methanosarcina mazei* Go1]	71 %	−3

Table 7.9 Sequences of the three differentially expressed fragments isolated from Colonial and Velvet bentgrasses grown in greenhouse.

Primers	Size(bp)	Species	Sequence
P6-T3	128	Colonial bentgrass	TAT TAA CCC TCA CTA AAT GCT GGG TGA TGC ACT GGC GCT CTA ATA TGC CTA TGC AAT CAG CTT CAT ATT TCT CTT TCT TAA CTT CTA CAT TCG GAC ACT AAA AAA AAA GAT ATC ACT CAG CAT AAT GA
P2-T5	327	Velvet bentgrass	TAT TAA CCC TCA CTA AAT CGG TTC ATA GTT GAG AGG GGA ATA GTA TAA CAT AGG AAG ACC CTT TAT CGA TAC TAA GAA CAA AAT GGT TTT TTT TGA TTG GAT TAG GAA TTC CCT CTT TTC TTT TTT TAA TCC TTT TCT CCT TTT TCT TTT CTA TAC CTC TAA CCC ACG AAT CTT TCT TAA TCT TAT CCA ATT TCC CTT ACT TTT TCT TAT AGT TAT ACA TAC AAT TAT GTA TGT ATT ATA TGA CCA ACT TTC TAT GGG TCA CAT AGA CAT CCA AAT AAG CAG TAG AAC TGA CAC GGG GAA AAA AAA AGA TAT CAC TCA GCA TAA TGA
P3-T8	146	Velvet bentgrass	TAT TAA CCC TCA CTA AAT GCT GGG TGG TTT TGG GCT ACA ATA GAT GAC GAG GGA ACT AGT TAT TAA GTG TGC TTG AAA CCT ATC CAT GGT ACA TTC ACA GAA TTG AGC GAC GGT GGC AAA AAA AAA GAT ATC ACT CAG CAT AAT GA

7.4 Discussion

The amount of Zn in seed is important when comparing the tolerance levels of different species to Zn deficiency. The Zn content of the seeds indicated that Velvet bentgrass seeds had the highest amount of Zn among the three species, whereas the other two had comparable contents of Zn in their seeds. According to the dry weights of plants after 4 weeks of growth under Zn deficiency, Velvet bentgrass was almost unresponsive to 0.1 ppm Zn application, which may be accounted by its high amount of Zn in the seeds. Despite the seed Zn contents, Creeping bentgrass was found to be the most tolerant species to Zn deficiency, compared to Colonial and Velvet bentgrasses, as indicated by its higher dry weight under varying Zn supplies. Conversely, the most susceptible species against Zn deficiency among the three was Velvet bentgrass. Highest dry weights of Creeping bentgrass was also reflected in the total Zn content per plant, although the Zn concentration of the shoots were generally lower compared to Colonial and Velvet bentgrasses (Fig. 7.2). The unexpected peak of Zn concentration in Colonial bentgrass in the absence of Zn supply (Fig. 7.2a) was concluded to derive from soil contamination before the ICP measurements rather than Zn contamination in the soil, since the same deviance was not observed for the dry weights of the plants.

The mRNADD method enabled identification of a total of nine Zn responsive transcripts, six in Creeping, one in Colonial, and two in Velvet bentgrass, that were differentially expressed under differing Zn supplies. The mRNADD experiments were repeated for several times and only the absolutely reproducible fragments have been analyzed further. The nonreproducible fragments were deemed unreliable and omitted. Majority of the identified transcripts revealed partial high similarities with the previously identified metal-binding proteins. These results clearly indicate that mRNADD approach efficiently identified Zn responsive genes, which can be mapped to molecular linkage maps to further utilize in molecular breeding programs for fortified grass crops.

Obtaining the full sequences of the transcripts will provide more information about the homology and orthology with the previously identified proteins. In addition, ORF regions of the full mRNAs can be used to translate the product in a model organism for structural and biochemical characterizations. The fragment isolated from Creeping bentgrass that was found to share extensive similarity with three ZIP transporters with approximately 50 % identity. Hence, the expression of this differentially expressed gene, a putative member of the ZIP family, was further analyzed. The quantitative real-time PCR results for the expression analysis of the fragment clearly indicated that its expression level was highly affected by the Zn concentration in the growth medium, evidenced by expression data from both tissue culture and greenhouse conditions. This is consistent with the previous findings showing that the expression of genes encoding the plant ZIP transporters appears to be affected by Zn or iron deficiencies (Eide et al. 1996; Grotz et al. 1998). The full-length mRNA sequence was found to be homologous to ZIP transporter family proteins which are mostly membrane proteins involved in transportation of the Zn after the uptake. Decrease in the expression level of this gene by decreasing amounts of Zn supply can be explained by its role, so that if Zn levels are high in the medium, the expression level of the gene increases for its transport into the plant.

The fragment that was obtained by P3-T1 primers had two specific ORFs and these were cloned to an expression vector and expressed in *Escherichia coli*. However, both protein products could not be obtained probably because they could not complete their proper folding, and become targets of protease activity. Consequently, the sequence of the full-length mRNA of the fragment was obtained by RACE. The full-length mRNA sequence, names as AsZIP, obtained by RACE was found to be a good candidate to be Creeping bentgrass homologous of ZIP family metal transporters detected in many organisms including plants, humans, and bacteria. This result was supported with BLASTX results and indicated that the protein product was highly similar to Zn transporter ZIP3 identified in *Oryza sativa* (OsZIP3, gbAAP85537.1) (Zhao and Eide 1996; Ramesh et al. 2003). According to Ramesh et al. (2003), the cDNA was partially complemented to a yeast mutant, deficient in Zn uptake at low-Zn concentrations. Considering that OsZIP3 was expressed under control conditions, but slightly upregulated when Zn was removed from the medium, they concluded that OsZIP3 is constitutively active and involved in overall cell Zn homeostasis, particularly in leaves.

The present results suggest that conserved mechanisms may play a role in tolerance against Zn deficiency in some *Agrostis* species. Further dissection of these mechanisms may reveal elusive species-specific patterns of tolerance against deficiency for better utilization of Zn, which breeding programs in related grass benefit from.

Acknowledgments BAA is a recipient of Tubitak-BİDEB scholarship.

References

Alloway B (2004) Zinc in soils and crop nutrition. In: Int. Zinc As- sociation Brussels, Belgium. http://zinc-crops.ionainteractive.com/publications/Zn_in_Soils_and_Crop_Nutrition_2008.pdf. Accessed 28 Jun 2014

Belanger FC, Meagher TR, Day PR, Plumley K, Meyer WA (2003) Interspecific Hybridization between and related species under field conditions. Crop Sci 43:240–246

Budak H, Su S, Ergen N (2006) Revealing constitutively expressed resistance genes in *Agrostis* species using PCR-based motif-directed RNA fingerprinting. Genet Res 88:165–175

Eide D, Broderius M, Fett J, Guerinot ML (1996) A novel iron-regulated metal transporter from plants identified by functional expression in yeast. Proc Natl Acad Sci U S A 93:5624–5628

Graham RD, Welch RM, Bouis HE (2001) Addressing micronutrient malnutrition through enhancing the nutritional quality of staple foods: Principles, perspectives and knowledge gaps. Adv Agron 70:77–142

Grotz N, Fox T, Connolly E, Park W, Guerinot ML, Eide D (1998) Identification of a family of zinc transporter genes from Arabidopsis that respond to zinc deficiency. Proc Natl Acad Sci 95:7220–7224

Hitchcock AS (1935) Manual of the grasses of the United States/ by A.S. Hitchcock. 1–1052

Kellogg EA (2001) Evolutionary History of the Grasses. PLANT Physiol 125:1198–1205

Liang P, Pardee AB (1992) Differential display of eukaryotic messenger RNA by means of the polymerase chain reaction. Science 257:967–971

López-Millán A-F, Ellis DR, Grusak MA (2004) Identification and characterization of several new members of the ZIP family of metal ion transporters in Medicago truncatula. Plant Mol Biol 54:583–596

Ramesh SA, Shin R, Eide DJ, Schachtman DP (2003) Differential metal selectivity and gene expression of two zinc transporters from rice. Plant Physiol 133:126–134

Shantz HL (1954) The Place of Grasslands in the Earth's Cover. Ecol 35(2):143–145

Van der Zaal BJ, Neuteboom LW, Pinas JE, Chardonnens AN, Schat H, Verkleij JA, Hooykaas PJ (1999) Overexpression of a novel Arabidopsis gene related to putative zinc-transporter genes from animals can lead to enhanced zinc resistance and accumulation. Plant Physiol 119:1047–1055

Zhao H, Eide D (1996) The ZRT2 gene encodes the low affinity zinc transporter in Saccharomyces cerevisiae. J Biol Chem 271:23203–23210

Chapter 8
Candidate Gene Approach in *Miscanthus* spp. for Biorefinery

Toshihiko Yamada, Hironori Nagano, Maria S. Dwiyanti, Lindsay V. Clark and Erik J. Sacks

8.1 Introduction

Global climate change and energy security have accelerated the interest in production and increased availability of alternative energy sources. Following increasing demand of sustainable-renewable energy sources, feedstock forages gain attention as bioenergy crop candidate. The advantages of perennial grass species as bioenergy crop are no annual establishment or tillage costs, low requirement for fertilizer and herbicides, currently free of serious diseases and pests, high biomass potential, lower interference with food crop production and good adaptation to marginal lands. Among perennial grass species, *Miscanthus* spp. and switchgrass (*Panicum virgatum* L.) are now the primary choices for bioenergy crops in temperate regions.

The genus *Miscanthus* is a rhizomatous perennial C_4 grass that grows naturally in east and southeastern Asia, Himalaya, southern Africa and Melanesia (Hodkinson et al. 2002). Taxonomically, *Miscanthus* belongs to Andropogoneae group with maize (*Zea mays* L.), sorghum (*Sorghum bicolor* (L.) Moench) and sugarcane (*Saccharum officinarum* L.) (Hodkinson et al. 2002). *Miscanthus* is composed of 14–20 species (Adati and Shiotani 1962; Sacks et al. 2013; Dwiyanti et al. 2013b). Among them, *M. sinensis* Anderss., *M. sacchariflorus* (Maxim.) Bentham and *M.* × *giganteus* Greef & Deuter ex Hodkinson & Renvoize gain attention as bioenergy crop candidate. *M. sinensis* is naturally distributed in east and southeastern Asia and it is typically diploid (Lafferty and Lelley 1994; Sacks et al. 2013). *M. sacchariflorus* is native to East Asia, and contains diploid and tetraploid plants. Most of diploid *M. sacchariflorus* can be found in China, whereas tetraploid is found in Japan. To

T. Yamada (✉) · H. Nagano · M. S. Dwiyanti
Field Science Center for Northern Biosphere, Hokkaido University,
Kita 11 Nishi 10, Sapporo, Hokkaido 060–0811, Japan
e-mail: yamada@fsc.hokudai.ac.jp

L. V. Clark · E. J. Sacks
Department of Crop Sciences, University of Illinois,
Urbana-Champaign, Urbana, IL 61801, USA

H. Budak, G. Spangenberg (eds.), *Molecular Breeding of Forage and Turf*,
DOI 10.1007/978-3-319-08714-6_8

date, commercial production of *Miscanthus* is limited to the cultivation of a single clone of *M.* ×*giganteus*; a sterile interspecific hybrid between *M. sinensis* and *M. sacchariflorus* which is brought from Yokohama, Japan in the 1930s by Aksel Olsen (Greef and Deuter 1993; Linde-Laursen 1993). Field trials have shown that *M.* ×*giganteus* is easy to grow, currently free of serious diseases and pests and capable of high-biomass productivity (BP) under cool temperatures (Clifton-Brown et al. 2008; Chou 2009). However, *M.* ×*giganteus* suffers from several limitations including pollen sterility due to its triploid nature, necessitating vegetative rhizome propagation with high-establishment costs (Tuck et al. 2006; Clifton-Brown and Lewandowski 2002), reduced first season overwintering survival, and poor adaptation to drought stress. In northern Europe, low temperatures limit cultivation of *Miscanthus* (Lewandowski et al. 2000; Danalatos et al. 2007), although once established it can yield up to 44 t ha^{-1} in northern America and 15–25 t ha^{-1} in Europe (Clifton-Brown et al. 2004; Zub and Brancourt-Hulmel 2010). Accordingly, scientists are seeking for new *Miscanthus* genetic resources with improved overwintering survival and high BP in cold environments.

Effort has been put to find more hybrids from natural populations resulting in discovery of three *M.* ×*giganteus* from sympatric population of *M. sinensis* and *M. sacchariflorus* in Kushima area, Japan (Nishiwaki et al. 2011; Dwiyanti et al. 2013a). In addition to exploring natural populations, restoring *M.* ×*giganteus* fertility by genome duplication (Yu et al. 2009) or by crossing *M. sacchariflorus* and *M. sinensis* has been attempted. *M. sinensis* also gains attention as bioenergy crop alternative to *M.* ×*giganteus,* because of higher cold tolerance and BP in cold region, the availability of genetic resources and possibility to develop new genotype through conventional breeding program (Stewart et al. 2009; Hodgson et al. 2010; Anzoua et al. in press).

One of the targets of improvement of *Miscanthus* is to change the content and optimize the ratio of lignin, cellulose and hemicelluloses in cell walls (Hisano et al. 2009). This is one of the most popular approaches for improvement of saccharification which is important for bioenergy crop because high saccharification benefits fermentation of sugars into ethanol (Li et al. 2008; Hisano et al. 2009). Postponing or eliminating flowering, modifying plant height, tiller number and density and stem thickness are considerable challenges for maximization of biomass yield (Jakob et al. 2009). In addition, improving the tolerance to biotic and abiotic stresses is necessary for stable BP. In this chapter, we outlined our recent research activities on molecular breeding such as candidate gene approach in *Miscanthus* spp.

8.2 Miscanthus Genetic Structure

Miscanthus species are native to temperate, subtropical and tropical areas of East Asia (Hodkinson et al. 2002). *M. sinensis* is the most broadly distributed in Asia and found from sea level to 2500 m primarily in China, Korea and Japan, but as far north as Sakhalin and as far south as the Indochinese peninsula (Lee 1964a, b; Sun et al. 2010; Sacks et al. 2013). *M. sinensis* is adapted primarily to environments that have an average annual minimum temperature of –28.9 ℃ or greater (USDA hardiness zone 5 or warmer) and receive ≥ 750 mm of precipitation annually (Clifton-Brown et al. 2008; Sun et al. 2010; Sacks et al. 2013). *M. sinensis* is an early colonizer after ecological disturbance in environments that would otherwise support forest (Numata 1969; Stewart et al. 2009). Wide geographical range suggests opportunities for isolation and differentiation of populations. However, *M. sinensis* is self-incompatible and has wind-dispersed pollen and seed, which are traits that would be expected to limit differentiation of populations. The degree of population differentiation using molecular markers has only been partially evaluated for *M. sinensis* over some of its native range, including studies in South Korea (Qin et al. 2013), Japan (Shimono et al. 2013), the Izu Islands of Japan (Iwata et al. 2005, 2006), Taiwan and the Ryukyu Islands (Chou et al. 2000), and China (Xu et al. 2013; Zhang et al. 2013; Zhao et al. 2013), and between South Korea and Japan (Slavov et al. 2013). *M. sinensis* in Japan were found to be distinct from those in Korea (Slavov et al. 2013), and structure was detected among populations in Japan (Shimono et al. 2013) and in China (Xu et al. 2013; Zhang et al. 2013; Zhao et al. 2013). Recently, many accessions of *M. sinensis* from most of its native range in China, Japan and South Korea were evaluated with restriction site-associated DNA sequencing single nucleotide polymorphism (SNP) markers, Golden Gate SNPs and ten plastid microsatellite markers (Clark et al. 2014). Six genetic clusters of *M. sinensis* from geographically distinct regions in Asia were identified. Clark et al. (2014) supposed that southeastern China is the origin of *M. sinensis* populations found in temperate eastern Asia, which is consistent with this area probably having been a refugium during the last glacial maximum (LGM) and speculated that *M. sinensis* migrated directly from southeastern China to Japan before migrating to the same latitudes in China and Korea, which is consistent with the known sequence of warming post-LGM. Combining phenotype data with genotype data in genome wide association study and the identification of potential heterotic groups could be useful for future *Miscanthus* breeding program.

8.3 Candidate Gene Approach

8.3.1 Candidate Gene Approach on Genes Involved in Lignin Biosynthesis Pathway

Lignin is a complex aromatic heteropolymers derived mainly from three types of monolignols differing in their degree or methoxylation, *p*-hydroxyphenyl (H), guaiacyl (G) and syringyl (S) (Boerjan et al. 2003; Vanholme et al. 2010). These monolignols composition largely varies depend on plant species, cell types, individual cell wall layers, developmental stage, and environment condition (Boerjan et al. 2003). Lignin forms complex with cellulose, which inhibits access of cellulase to cellulose during hydrolysis process. Therefore, bioenergy crop with lower lignin content is desirable for reducing hydrolysis cost in biorefinery process. Lignin subunits are synthesized from phenylalanine, and its biosynthesis pathway has been extensively studied in gymnosperm and angiosperm (Boerjan et al. 2003; Vanholme et al. 2010). Numerous lignin biosynthesis genes have been cloned and studied in various species, such as Arabidopsis, maize, alfalfa, tobacco, poplar and rice (Boerjan et al. 2003; Vanholme et al. 2010).

As preliminary study on lignin biosynthesis genes of *Miscanthus* species, genes encoding cinnamyl alcohol dehydrogenase (*CAD*), caffeic *O*-methyltransferase (*COMT*) and 4-coumarate: coenzyme A ligase (*4CL*) from *M. sinensis* and *M. sacchariflorus* were cloned (Dwiyanti et al. 2014). Primers used in gene cloning were designed based on genes mutated in *bmr* mutants which are natural low lignin mutants. The genes are *Sb04g005950*, *Sb07g003860*, and *Sb04g005210* for CAD, COMT, and 4CL, respectively. Full length cDNA was cloned for *CAD* gene, whereas partial cDNA was cloned for *COMT* and *4CL*. Based on sequence analysis, three copies of *CAD* gene corresponding to *Sb04g005950* were identified for both *M. sinensis* and *M. sacchariflorus*. Three and five copies of *COMT* gene corresponding to *Sb07g003860* were identified for *M. sinensis* and *M. sacchariflorus*, respectively; whereas four copies of *4CL* gene corresponding to *Sb04g005210* were identified for both *M. sinensis* and *M. sacchariflorus*. Based on the amino acid sequence similarity the *4CL* and *CAD* genes can be grouped into type I and type II. Expression level of the three genes on different stem parts (upper and lower) before flowering time (July) were analyzed for a *M. sinensis* accession (Shiozuka), a *M. sacchariflorus* accession (Miyakonojo), *M.* × *giganteus* (Illinois), and a triploid hybrid between *M. sacchariflorus* and *M. sinensis* found by Nishiwaki et al. (2011). The expression analysis showed that before flowering time, the genes were expressed higher in lower stem than in upper stem. Interestingly, at before flowering stage, CAD type I expression was higher than CAD type II in *M. sinensis* at upper and lower stem. Expression analyses at several different growing stages are essential to understand on how the gene expression changes over time. In addition, analysis of content and composition of lignin among *M. sinensis*, *M. sacchariflorus* and *M.* × *giganteus* is needed to understand the relationship between gene expression and change in lignin content/composition in *Miscanthus* species.

8.3.2 Candidate Gene Approach on Genes Involved in Control of Flowering Time

Flowering time is a key target trait for extending the vegetative phase to increase biomass potential. Gene regulation networks of flowering time under long-day condition were well identified in sorghum (Murphy et al. 2011) and rice (Koo et al. 2013). *CONSTANS* (*CO*)/ *Heading Date 1* (*Hd1*) is one of the central regulators for flowering pathway in Arabidopsis and rice, respectively (Valverde 2011). To obtain insight into the control of flowering time for *M. sinensis*, we cloned homologs of *Hd1*, which we call *MsiHd1* and compared among 18 wild-collected accessions of

M. sinensis from Japan and 13 from mainland Asia. Two to five *MsiHd1* alleles in each accession were identified, suggesting that *MsiHd1* consists of multiple loci in the *Miscanthus* genome. The first duplication generated *Hd1a* and *Hd1b*, and then *Hd1a* was further duplicated to more than two copies (*Hd1a1–Hd1a3*). Both of *Hd1a* and *Hd1b* had kept functional in *M. sacchariflorus*, while *Hd1b* in *M. sinensis* and *M. floridulus* had pseudogenize. Accessions of the mainland Asia in *Hd1a* had only functional alleles, on the other hand, accessions of Japanese islands preserved many nonfunctional alleles, which were caused by indel and miniature inverted transposable elements (MITE). Further studies on identification and characterization of other flowering genes and association of genotypes with flowering phenotypes are need to understand gene regulation networks of flowering time in *Miscanthus*.

8.4 Genetic Transformation

The genetic transformation method for *Miscanthus* is still difficult, because at present, there is only a few detailed report of success in *Miscanthus* (Wang et al. 2011; Hwang et al. 2014). Particle bombardment mediated transformation systems were used to transform *M. sinensis* with perennial ryegrass fructosyltransferase (*prft*) genes enabling transgenic plants to produce fructans. Fructans were detected in transgenic plants expressing *prft4* gene, encoding sucrose-sucrose 1-fructosyltransferase (1-SST). The transgenic plants showed minor chlorophyll loss and fewer leaves wilting under cold temperature conditions. This indicated that the transgenic plants are expected to produce higher photosynthetic biomass under low temperature (unpublished).

8.5 Conclusion

High biomass yield with low cost of *Miscanthus* makes it promising as biomass crop, however, *Miscanthus* is a new crop that needs development. Most breeding programs on *Miscanthus* are still at the initial phases. Knowledge on identification, characterization and function of genes involved with BP in *Miscanthus* is still in shortage. The perennial habit and complexity of the genome are limiting factors in elucidating the genetic basis of *Miscanthus* agronomical traits and quality traits. Recently, Swaminathan et al. (2012) analyzed deep transcriptome sequencing (RNAseq) from two *M. sinensis* accessions to define 1536 single nucleotide variants (SNVs) for a GoldenGate™ genotyping array and contrasted a linkage map of *M. sinensis*. A high-resolution linkage map of *M. sinensis* was also created using genotyping-by-sequencing (GBS; Ma et al. 2012). The recent advances in "-omics" technologies may accelerate the progress of whole genome sequencing, genetic marker development and elucidation of physiological process in *Miscanthus*.

References

Adati S, Shiotani I (1962) The cytotaxonomy of the genus *Miscanthus* and its phylogenic status. Bull Fac Agric Mie Univ 25:1–14

Anzoua KG, Suzuki K, Fujita S, Toma Y, Yamada T (2015) Evaluation of morphological traits, winter survival and biomass potential in wild Japanese *Miscanthus sinensis* Anderss. Populations in Northern Japan. Grassland Science, in press

Boerjan W, de Andrade MG, Mazzafera P (2003) Lignin biosynthesis. Ann Rev Plant Biol 54:519–546

Chou CH (2009) *Miscanthus* plants used as an alternative biofuel material: the basic studies on ecology and molecular evolution. Renew Energy 34:1908–1912

Chou CH, Chiang YC, Chiang TY (2000) Genetic variability and phytogeography of *Miscanthus sinensis* var. *condensatus*, an apomictic grass, based on RAPD fingerprints. Can J Bot 78:1262–1268

Clark LV, Brummer JE, Głowacka K, Hall M, Heo K, Long SP, Peng J, Yamada T, Yoo JH, Yu CY, Zhao H, Sacks EJ (2014) A footprint of past global climate change on the population genetic structure of *Miscanthus sinensis*. Ann Bot, 141:97–107

Clifton-Brown JC, Lewandowski I (2002) Screening *Miscanthus* genotypes in field trials to optimise biomass yield and quality in Southern Germany. Eur J Agron 16:97–110

Clifton-Brown JC, Stampfl PF, Jones MB (2004) *Miscanthus* biomass production for energy in Europe and its potential contribution to decreasing fossil fuel carbon emission. Glob Change Biol 10:509–518

Clifton-Brown J, Chiang Y-C, Hodkinson TR (2008) *Miscanthus*: genetic resources and breeding potential to enhance bioenergy production. In: Vermerris W (ed) Genetic improvement of bioenergy crops. Springer, Berlin, pp 273–294

Danalatos NG, Archontoulis SV, Mitsios I (2007) Potential growth and biomass productivity of *Miscanthus* × *giganteus* as affected by plant density and N-fertilization in central Greece. Biomass Bioenergy 31:145–152

Dwiyanti MS, Rudolph A, Swaminathan K, Nishiwaki A, Shimono Y, Kuwabara S, Matuura H, Nadir M, Moose S, Stewart JR, Yamada T (2013a) Genetic analysis of putative triploid *Miscanthus* hybrids and tetraploid *M. sacchariflorus* collected from sympatric populations of Kushima, Japan. BioEnergy Res 6:486–493

Dwiyanti MS, Stewart JR, Yamada T (2013b) Germplasm resources of *Miscanthus* and their application in breeding. In: Saha MC, Bhandari HS, Bouton JH (eds) Bioenergy Feedstocks: breeding and Genetics. Wiley-Blackwell, Ames pp 49–66

Dwiyanti MS, Stewart JR, Yamada T (2014) Forages for feedstocks of biorefineries in temperate environments: review of lignin research in bioenergy crops and some insight into *Miscanthus* studies. Crop & Pasture Science, 65:1199–1206

Greef JM, Deuter M (1993) Syntaxonomy of *Miscanthus* × *giganteus* Greef et Deu. Angew Bot 67:87–90

Hisano H, Nandakumar R, Wang ZY (2009) Genetic modification of lignin biosynthesis for improved biofuel production. In Vitro Cell Dev Biol Plant 45:306–313.

Hodgson EM, Lister SJ, Bridgwater AV, Clifton-Brown J, Donnison IS (2010) Genotypic and environmentally derived variation in the cell wall composition of *Miscanthus* in relation to its use as a biomass feedstock. Biomass Bioenergy 34:652–660

Hodkinson TR, Chase MW, Lledó MD, Salamin N, Renvoize SA (2002) Phylogenetics of *Miscanthus*, *Saccharum* and related genera (Saccharinae, Andropogoneae, Poaceae) based on DNA sequences from ITS nuclear ribosomal DNA and plastid *trnL* intron and *trnL-F* intergenic spacers. J Plant Res 115:381–392

Hwang OJ, Cho MA, Han YJ, Kim YM, Lim SH, Kim DS, Hwang I, Kim JI (2014) *Agrobacterium*—mediated genetic transformation of *Miscanthus sinensis*. Plant Cell Tiss Organ Cult 117:51–63

Iwata H, Kamijo T, Tsumura Y (2005) Genetic structure of *Miscanthus sinensis* ssp. *condensatus* (Poaceae) on Miyake Island: implications for revegetation of volcanically devastated sites. Ecol Res 20:233–238

Iwata H, Kamijo T, Tsumura Y (2006) Assessment of genetic diversity of native species in Izu Islands for a discriminate choice of source populations: implications for revegetation of volcanically devastated sites. Conserv Genet 7:399–413

Jakob K, Zhou F, Paterson AH (2009) Genetic improvement of C_4 grasses as cellulosic biofuel feedstocks. In Vitro Cell Dev Biol Plant 45:291–305

Koo BH, Yoo SC, Park JW, Kwon CT, Lee BD, Gynheung A, Zhang Z, Li J, Li Z, Paek NC (2013) Natural variation in *OsPRR37* regulates heading date and contributes to rice cultivation at a wide range of latitudes. Mol Plant 6:1877–1888

Lafferty J, Lelley T (1994) Cytogenetic studies of different *Miscanthus* species with potential for agricultural use. Plant Breed 113:246–249

Lee Y (1964a) Taxonomic studies of the genus *Miscanthus*: relationships among the section, subsection, and species, part 1. J Jpn Bot 39:196–205

Lee Y (1964b) Taxonomic studies on the genus *Miscanthus*: relationships among the section, subsection, and species, part 2: enumeration of species and varieties. J Jpn Bot 39:257–265

Lewandowski I, Clifton-Brown JC, Scurlock JMO, Huisman W (2000) *Miscanthus*: European experience with a novel energy crop. Biomass Bioenergy 19:209–227

Li X, Weng J-K, Chapple C (2008) Improvement of biomass through lignin modification. Plant J 54:569–581

Linde-Laursen I (1993) Cytogenetic analysis of *Miscanthus* "Giganteus", an interspecific hybrid. Hereditas 119:297–300

Ma X-F, Jensen E, Alexandrov N, Troukhan M, Zhang L, Thomas-Jones S, Farrar K, Clifton-Brown J, Donnison I, Swaller T, Flavell R (2012) High resolution genetic mapping by genome sequencing reveals genome duplication and tetraploid genetic structure of the diploid *Miscanthus sinensis*. PLoS ONE 7:e33821

Murphy RL, Klein RR, Morishige DT, Brady JA, Rooney WL, Miller FR, Dugas DV, Klein PE, Mullet JE (2011) Coincident light and clock regulation of pseudoresponse regulator protein 37 (PRR37) controls photoperiodic flowering in sorghum. Proc Natl Acad Sci USA 108:469–474

Nishiwaki A, Mizuguti A, Kuwabara S, Toma Y, Ishigaki G, Miyashita T, Yamada T, Matuura H, Yamaguchi S, Rayburn AL, Akashi R, Stewart JR (2011) Discovery of natural *Miscanthus* (Poaceae) triploid plants in sympatric populations of *Miscanthus sacchariflorus* and *Miscanthus sinensis* in southern Japan. Amer J Bot 98:154–159

Numata M (1969) Progressive and retrogressive gradient of grassland vegetation measured by degree of succession—ecological judgment of grassland condition and trend IV. Plant Ecol 19:96–127

Qin Y, Kabir MA, Wang HW et al (2013) Assessment of genetic diversity and relationships based on RAPD and AFLP analyses in *Miscanthus* genera landraces. Can J Plant Sci 93:1–12

Sacks EJ, Juvik JA, Lin Q, Stewart JR, Yamada T (2013) The gene pool of *Miscanthus* species and its improvement. In: Paterson AH (ed) Genomics of the Saccharinae. Springer, New York, pp 73–101

Shimono Y, Kurokawa S, Nishida T, Ikeda H, Futagami N (2013) Phylogeography based on intraspecific sequence variation in chloroplast DNA of *Miscanthus sinensis* (Poaceae), a native pioneer grass in Japan. Botany 91:449–456

Slavov G, Robson P, Jensen E et al (2013) Contrasting geographic patterns of genetic variation for molecular markers vs. phenotypic traits in the energy grass *Miscanthus sinensis*. GCB Bioenergy 5:562–571

Stewart JR, Toma YO, Fernandez FG, Nishiwaki A, Yamada T, Bollero GN (2009) The ecology and agronomy of *Miscanthus sinensis* a species important to bioenergy crop development, in its native range in Japan: a review. GCB Bioenergy 1:126–153

Sun Q, Lin Q, Yi Z-L, Yang Z-R, Zhou F-S (2010) A taxonomic revision of *Miscanthus* s.l (Poaceae) from China. Bot J Linnean Soc 164:178–220

Swaminathan K, Chae WB, Mitros T, Varala K, Xie L, Barling A, Glowackal K, Hall M, Jezowski S, Ming R, Matthew Hudson M, Juvik JA, Rokhsar DS, Moose SP (2012) A framework genetic map for *Miscanthus sinensis* from RNAseq-based markers shows recent tetraploidy. BMC Genom 13:142

Tuck G, Glendining MJ, Smith P, House JI, Wattenbach M (2006) The potential distribution of bioenergy crops in Europe under present and future climate. Biomass Bioenergy 30:183–197

Valverde F (2011) CONSTANS and the evolutionary origin of photoperiodic timing of flowering. J Exp Bot 62:2453–2463

Vanholme R, Demedts B, Morreel K, Ralph J, Boerjan W (2010) Lignin biosynthesis and structure. Plant Physiol 153:895–905

Wang X, Yamada T, Kong F-J, Abe Y, Hoshino Y, Sato H, Takamizo T, Kanazawa A, Yamada T (2011) Establishment of an efficient in vitro culture and particle bombardment-mediated transformation systems in *Miscanthus sinensis* Anderss., a potential bioenergy crop. GCB Bioenergy 3:322–332

Xu W-Z, Zhang X-Q, Huang L-K, Nie G, Wang J-P (2013) Higher genetic diversity and gene flow in wild populations of *Miscanthus sinensis* in southwest China. Biochem Syst Ecol 48:174–181

Yu CY, Kim HS, Rayburn AL, Widholm JM, Juvik JA (2009) Chromosome doubling of the bioenergy crop, *Miscanthus × giganteus*. GCB Bioenergy 1:404–412

Zhang QX, Shen YK, Shao RX et al (2013) Genetic diversity of natural *Miscanthus sinensis* populations in China revealed by ISSR markers. Biochem Syst Ecol 48:248–256

Zhao H, Wang B, He J, Yang J, Pan L, Sun D, Peng J (2013) Genetic diversity and population structure of *Miscanthus sinensis* germplasm in China. PLoS ONE 8:e75672

Zub HW, Brancourt-Hulmel M (2010) Agronomic and physiological performances of different species of *Miscanthus*, an important energy crop. A review. Agron Sustain Dev 30:201–214

Chapter 9
Genetic Diversity in Candidate Genes for Developmental Traits and Cell Wall Characteristics in Perennial Ryegrass (*Lolium perenne*)

T. Ruttink, A. Haegeman, F. van Parijs, S. Van Glabeke, H. Muylle, S. Byrne, T. Asp and I. Roldán-Ruiz

9.1 Introduction

In outbreeding species in which linkage disequilibrium (LD) spans short distances, the use of markers in candidate genes is an appropriate approach for association mapping (AM) (Zhu et al. 2008). Several studies have successfully used this approach in *Lolium perenne*. For example, Skøt et al. (2007) examined the association between an alkaline invertase candidate gene (*LpcAI*) and carbohydrate content, and between the photoperiod control gene *LpHD1* and flowering time; Skøt et al. (2011) investigated the association between polymorphisms in the *L. perenne* ortholog of the *Arabidopsis* FT gene (*LpFT3*) and flowering time; Auzanneau et al. (2007) examined the association between *L. perenne Gibberelic Acid Insensitive* (*LpGAI*) and organ growth. Recently, Yu et al. (2013) examined the association between polymorphisms in a set of 14 genes and drought tolerance. However, these studies focused on a relatively small number of genes, which is, at least partially, due to the relatively large population sizes typical of AM studies and the lack of cost-efficient genotyping technologies for highly polymorphic genomes. These studies used either amplicon resequencing across the population (Auzanneau et al. 2007; Yu et al. 2013), or identified alleles in a representative discovery panel followed by screening the AM population with the subset of polymorphisms that discriminated those alleles using TaqMan assays (Skøt et al. 2007) or PCR-fragment length polymorphisms (Skøt et al. 2011) as genotyping tool. While analysis of a limited number of candidate genes can be appropriate for the dissection of traits with a clearly defined genetic control and easy to pin-point candidate genes, it represents a

T. Ruttink (✉) · A. Haegeman · F. van Parijs · S. Van Glabeke · H. Muylle · I. Roldán-Ruiz
Plant Sciences Unit—Growth and Development, Institute for Agricultural and Fisheries Research (ILVO), Caritasstraat 21, 9090 Melle, Belgium
e-mail: tom.ruttink@ilvo.vlaanderen.be

S. Byrne · T. Asp
Molecular Biology and Genetics, Aarhus University, Forsøgsvej 1, 4200 Slagelse, Denmark

H. Budak, G. Spangenberg (eds.), *Molecular Breeding of Forage and Turf*,
DOI 10.1007/978-3-319-08714-6_9

clear limitation when dealing with complex traits. Indeed, even when a significant association is found between polymorphisms in one candidate gene and the investigated trait, it is difficult to rule out that other genes might also have an effect, or that (part of) the phenotypic diversity observed is due to interactions between multiple genes. Furthermore, most high-throughput genotyping platforms only allow screening of *a priori* known SNPs, so that only a subset of the potentially relevant genetic diversity is considered in the AM study. As a consequence, the development of versatile, cost-efficient methods to screen a large number of polymorphisms in multiple candidate genes for a diverse set of *L. perenne* genotypes is required. Targeted resequencing of a panel of several hundred candidate genes using next generation sequencing (NGS) provides the possibility to simultaneously discover and genotype *a priori* unknown and/or low-abundant polymorphisms across a large population.

We have investigated whether probe capture enrichment followed by NGS is a feasible approach for targeted resequencing in the highly heterozygous species *L. perenne*, for which high SNP densities of about five SNPs per 100 bp have been reported (Yu et al. 2013; Ruttink et al. 2013). In this approach, oligonucleotide baits are designed to bind regions of interest (i.e. in candidate genes), which are enriched by in-solution sequence capture before sequencing (Davey et al. 2011). Our final objective is to conduct candidate gene AM of plant developmental traits and cell wall characteristics in a population of about 700 genotypes including breeding materials and wild accessions. Therefore, we identified candidate genes putatively involved in the regulation of plant growth and development, plant architecture, meristem activity, induction of flowering, cell wall biogenesis, and phytohormone biosynthesis, signaling, and response, and selected 539 *L. perenne* candidate genes for genotyping. During probe design, we excluded target regions containing repeat sequences or with high sequence similarity to paralogous genes in the reference genome sequence, which can affect target-probe hybridization during the enrichment step and cause ambiguous read mapping after sequencing (Uitdewilligen et al. 2013).

Here, we evaluate the performance of this newly developed probe set for the screening of 539 *L. perenne* genes, discuss the effect of read depth (RD) on target coverage, SNP and InDel density, and calling accuracy and evaluate LD patterns in this set of genes. The analysis is based on a set of 95 *L. perenne* genotypes from 17 accessions originating from France and the inbred line used for genome sequencing (Byrne et al. 2014).

9.2 Materials and Methods

9.2.1 Candidate Gene Identification and Probe Design

A total of 336 *Arabidopsis* genes involved in the regulation of plant growth and development, plant architecture, induction of flowering, cell wall biogenesis, and phytohormone biosynthesis, signaling, and response were selected based on literature searches. These 336 *Arabidopsis* genes belong to 177 different gene families according to the comparative genomics resource PLAZA2.5 (Van Bel et al. 2012).

All the corresponding *Brachypodium*, rice, maize, sorghum, and *Arabidopsis* proteins from these families were retrieved from PLAZA2.5. From these, the *Brachypodium* proteins were used to search (BLASTX) for all corresponding predicted gene models on the *L. perenne* draft genome sequence (Byrne et al. 2014), and these proteins were added to their respective protein family. A phylogenetic tree was constructed for each family and manually annotated to identify putative *L. perenne* orthologs for the 336 *Arabidopsis* candidate genes. If necessary, the predicted gene models were adjusted using RNA-seq transcriptome sequence alignments (Ruttink et al. 2013). A total of 539 *L. perenne* candidate genes were chosen for probe design, yielding an average length of 5168 bp target region per gene, or a total of 2.7 Mbp. The target regions on the draft genome sequence were defined as the predicted gene model of the candidate genes plus an additional 1000 bp upstream region, or until an upstream coding region was encountered.

OligoTiler (Bertone et al. 2006) was used to tile the forward (coding) reference strand of each continuous target region with 120 bp probes starting approximately every 40 bp. From this probe set, we eliminated probes with high sequence similarity to non-targets, repetitive sequences, or mitochondrial or chloroplast sequences. Finally, a total of 57,700 120 bp SureSelect probes (Agilent) were selected that cover 2.4 Mbp of the intended 2.7 Mbp target regions, at around 3× tiling.

9.2.2 Library Construction and Sequencing

Genomic DNA was extracted from 95 *L. perenne* genotypes from 17 accessions originating from France (inset in Fig. 9.5) and the inbred line used for genome sequencing (Byrne et al. 2014). For each genotype, we prepared an indexed shotgun library by (i) adaptive focused acoustic (AFA) fragmentation on a Covaris S2 instrument (Covaris, Inc.), (ii) adapter ligation, and (iii) magnetic bead purification, using a slightly adapted version of the protocol of Uitdewilligen et al. (2013). The 96 libraries were then pooled without normalization and used for a single probe capture hybridization reaction according to the SureSelect protocol (Agilent SureSelect^{XT2} Target Enrichment System for Illumina Paired-End Sequencing Library Protocol, v. 1.0). After probe capture, the purified enriched pooled library was split in two halves, and both parts were amplified in an independent PCR reaction rendering two libraries for sequencing. Each library was then sequenced on one lane of a HiSeq2000 using 91 bp paired-end sequencing, yielding 383 M and 336 M reads.

9.2.3 Quality Filtering and Read Mapping

The raw reads were trimmed and quality filtered using Trimmomatic v. 0.32 (Bolger et al. 2014). All scaffolds of the draft genome sequence that had a significant tBLASTn hit ($E < 1e^{-10}$) with any of the *Brachypodium* proteins of the 177 gene families were used to create a reference sequence for read mapping. The trimmed and filtered reads were mapped onto this reference using the BWA-MEM aligner of the software BWA

v. 0.7.4 (Li and Durbin 2010). Duplicate reads were marked using Picard-tools 1.100 (http://picard.sourceforge.net). The resulting BAM files were further processed using the Genome Analysis Toolkit (GATK) v. 2.7-4 (McKenna et al. 2010). Regions with insertions and deletions (InDels) were identified and locally realigned. An initial round of variant calling was performed using GATK's HaplotypeCaller and high-quality SNPs and InDels were extracted. These high-quality variants were used to recalibrate the base quality scores in the mapping files per sequencing lane. The mapping files were further processed in CLC Genomics Workbench v7.0.0, and with the DepthOf-Coverage program in GATK to calculate RD and coverage of the target regions.

Throughout this manuscript we use the term 'sequencing depth' to indicate the number of reads per genotype in one of the libraries or when the reads of both libraries were combined. In addition, we use the term 'read depth' to indicate the number of aligned reads (after unique read mapping and deduplication) on a given position of the target region. The median RD is calculated across all target regions and samples (expressed as x per position). The median RD is therefore a global measure, and it is possible that at certain positions the actual RD is lower, or even completely absent. We use the term 'coverage' to indicate the percentage of a target region length to which reads are aligned. The coverage is calculated as the number of positions with a minimal number of aligned reads at that specific position (RD threshold), divided by the total length of the intended target region in bp, and expressed as percentage. The coverage is indexed with the RD threshold (e.g. $coverage_{RD\geq1}$). Please note the discrimination in the use of the terms median 'read depth' which is a global measure across the target region, and 'RD threshold' which is a bp position-specific measure, only used in combination with the coverage parameter.

9.2.4 Sequence Variant Calling

The recalibrated mapping files were used as input for the final variant calling by GATK's HaplotypeCaller. Resulting SNPs and InDels were quality filtered using several criteria, for example only variants with a quality score higher than 30 and a quality by depth score above 2.0 were retained. Further processing of the VCF files containing variants was done with VCFtools 1.11 (Danecek et al. 2011) and custom Perl scripts. The inbreeding coefficient (F) was estimated for each individual using a method of moments in VCFtools using the option-het. The Identity-By-State between technical replicates of samples was determined using the quality filtered SNP dataset with the R-package SNPRelate v. 0.9.18 (Zheng et al. 2012).

9.2.5 Population Structure and LD Estimates

A phylogenetic tree comparing the SNP profiles of the 96 genotypes was calculated in SNPhylo v. 20140430 using 100 bootstraps and without filtering based on LD (option-l 1.0) (Lee et al. 2014). iTol (Letunic and Bork 2011) was used for graphical representation.

The LD between SNPs within each of the 539 target regions was estimated for two subpopulations using VCFtools 1.11. Subpopulation 1 consisted of 56 wild genotypes from all over France, and comprised a random subset of eight genotypes from accession 10370 originating from Nord-Pas-de-Calais (inset Fig. 9.5). Subpopulation 2 was a set of 41 genotypes from the 10370 accession. For these calculations we only retained pair-wise comparisons between SNPs with a minimum genotype quality (GQ) score of 20, a maximum of 20 % missing data per subpopulation and a minor allele frequency (MAF) of either 0, 5, or 10 % (see results and discussion). For each target region at each MAF the set of r^2 values was fitted to the common logarithm of the distance (in bp) between SNP pairs, using linear regression in R. The regression equation was then used to predict the base pair distance for which r^2 took the arbitrarily chosen value of 0.2. Only target regions with at least 100 pair-wise comparisons were considered for LD decay estimations.

9.3 Results and Discussion

9.3.1 Coverage Analysis

Targeted resequencing is based on shotgun sequencing libraries and covering the complete target region depends on random sampling effects. Therefore, a certain degree of oversampling (here referred to as the sequencing depth) is required to cover all positions of the intended target region at least once (Wendl 2006). In addition, for accurate genotype calling and to account for sequencing errors in NGS data, it is important that each position of the intended target region is sequenced several times, here referred to as the minimal RD threshold *per position*. A non-linear relationship exists between the sequencing depth and the percentage of the intended target sequence length covered at a given RD threshold (e.g. $\text{coverage}_{RD\geq 1}$ or $\text{coverage}_{RD\geq 8}$) per position (Fig. 9.1). As it is not trivial to theoretically predict how many reads are required per sample, we experimentally determined this relationship to create coverage saturation curves at various RD thresholds. Genotype calling saturation curves are discussed further below.

We pooled 96 libraries without normalization, providing us with a range of sequencing depths across the 96 genotypes. This pool was sequenced in duplicate. When the reads of both sequencing lanes were combined per sample, a median of 479 k reads were uniquely mapped on-target per genotype after duplicate removal, corresponding with a median RD of 12.6× per position (Fig. 9.1, box-plot on *X*-axis). Figure 9.1 shows that, in our experimental set-up, a median target region $\text{coverage}_{RD\geq 1}$ of 88 %, a $\text{coverage}_{RD\geq 6}$ of 71 %, and a $\text{coverage}_{RD\geq 8}$ of 64 % was achieved across all genotypes and target regions (box-plots on the *Y*-axis of Fig. 9.1). The $\text{coverage}_{RD\geq 1}$ saturation curve illustrates that >700 k uniquely mapped reads per genotype (≈ median RD of 15 ×) are required to reach the $\text{coverage}_{RD\geq 1}$ plateau, which lies at 90 % of the intended target region length of 2.7 Mbp in our study. These curves further suggest that about 2 million uniquely mapped reads per genotype are

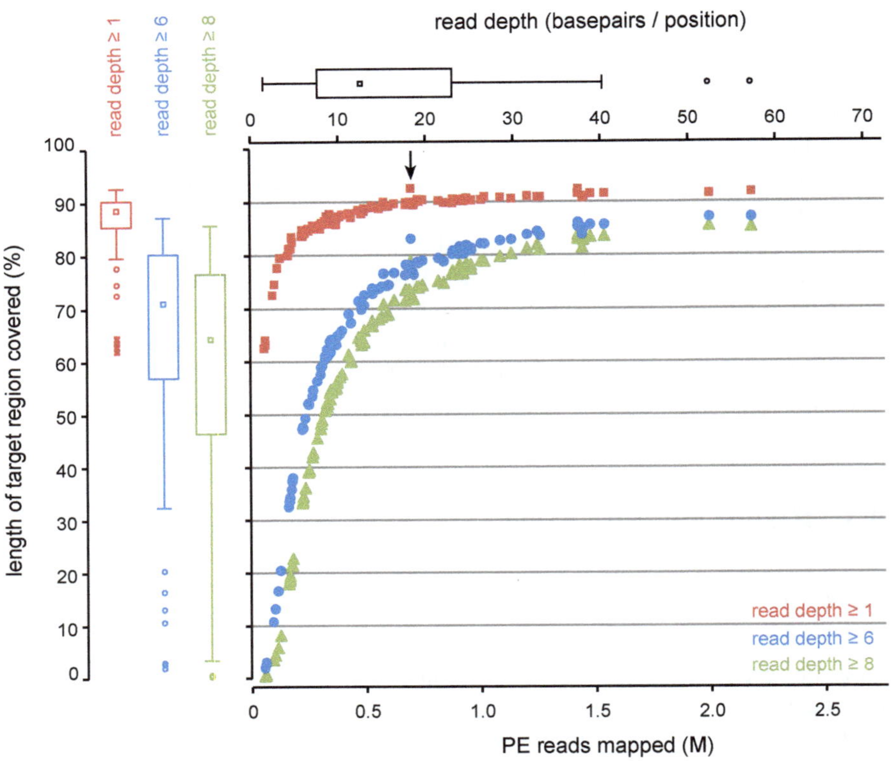

Fig. 9.1 Success rate of probe capture per genotype. The length of the target region covered is represented in function of the number of uniquely mapped reads on-target for each of the 96 genotypes, for different read depth (RD) thresholds (RD≥1, RD≥6, RD≥8). The *box-plots* on the *X*-axis represents the spread of the median RD across the 96 genotypes. The *box-plots* on the *Y*-axis represent the spread of the target length coverage at various RD thresholds. Per genotype the median coverage is calculated across all 539 target regions. The *arrow* indicates the reference genotype

required to reach coverage$_{RD\geq 8}$ of 85%. This can for instance be achieved with equal pooling of 96 samples but four times more sequencing data per pool.

Next, we investigated the resequencing success rate across the 539 target regions. The effect of the following three factors was investigated:

Probe design. If the capture probes bind repetitive regions, a disproportional enrichment of these sequences is obtained. In addition, repetitive sequences typically do not map uniquely and are excluded from SNP analysis, thus leading to loss of a portion of the sequencing capacity. Therefore, probes with high sequence similarity to non-targets, repetitive regions, chloroplast, and mitochondrial sequences (as detected by BLAST analyses) were removed during probe selection, except for a few genes for which sequencing was attempted despite high sequence similarity between closely related paralogs. The purple line in Fig. 9.2 shows the probe coverage per target region. A median of 90% of the intended length per target region was covered by the probes designed using our in-house developed custom procedure,

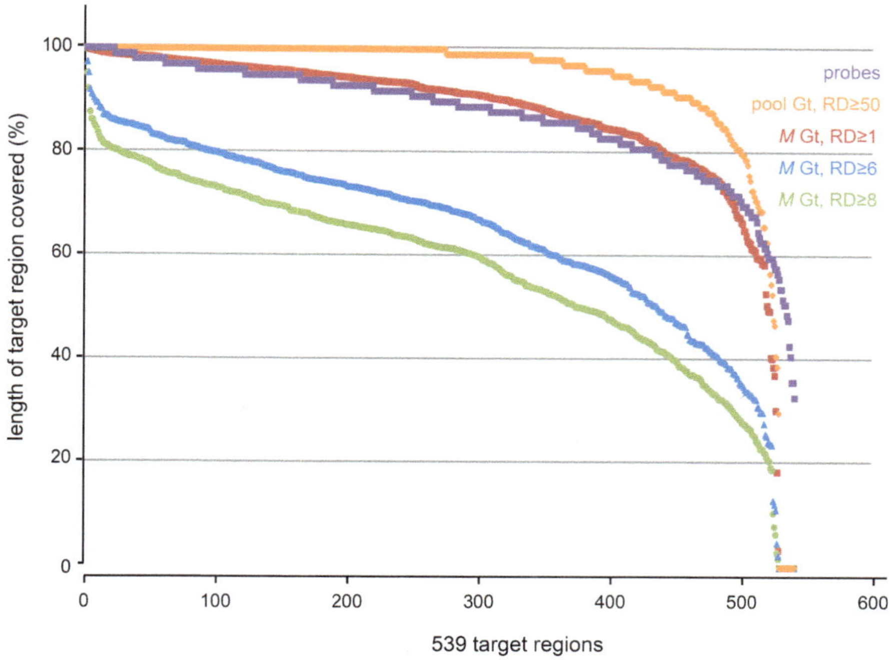

Fig. 9.2 Success rate of probe capture per target region. *Purple blocks*: coverage of the target region by probes. *Orange diamonds*: the coverage$_{RD\geq 50}$ of the target regions with the reads of all genotypes pooled (pool Gt). The *red squares* (coverage$_{RD\geq 1}$), *blue triangles* (coverage$_{RD\geq 6}$), and *green circles* (coverage$_{RD\geq 8}$): median coverage of the target region across all genotypes (*M* Gt) at different RD thresholds. Target regions are sorted separately per data series according to decreasing target region coverage

showing a fairly high success rate for probe design in genic regions of the *L. perenne* genome sequence.

Sequencing of regions flanking the probe targets. Given that a 120 bp probe can capture a 100–600 bp genomic DNA fragment in the shotgun library, a few hundred bases of flanking sequences can potentially be sequenced. This can compensate for small gaps in the probe coverage of the target regions. We used the pooled data of all 96 genotypes together for read mapping to estimate the fraction of each target region that was covered by sequence reads by this probe set. For 436 of 539 target regions, >95 % of the intended target region was effectively resequenced (coverage$_{RD\geq 50}$ in the pool; orange diamonds in Fig. 9.2), which is substantially higher than the fraction covered by probes (purple line in Fig. 9.2). For 12 target regions the RD was severely affected by non-unique mapping as expected (see above). Detailed analysis revealed that in all 12 cases the corresponding sequences were indeed captured. However, two or more highly similar sequences, e.g. closely related paralogs, tandem duplicated genes, or redundancy in the reference assembly, caused non-unique mapping of the reads and exclusion from RD and/or SNP analyses. This shows that removing probes targeting repetitive regions improves the information content of the resequencing data, or, alternatively, that local adjustment of the reference sequence may be required for optimal mapping.

Read depth. Due to the random sampling effects of shotgun libraries, a minimal number of reads *per sample* is required to cover the full length of the target region with sufficient RD for accurate SNP calling (Wendl 2006). To optimize the number of genotypes per hybridization/sequencing pool, it is important to estimate the minimal sequencing depth for a given degree of coverage and genotype calling on a per-sample level. Based on these coverage and/or genotype calling saturation curves, the experimental set-up can be shifted between sequencing more genotypes or more complete coverage. For this, we first determined the coverage per target region per genotype at a given RD threshold. We then calculated the median coverage per target region (across all genotypes) to estimate the global effective success rate of targeted resequencing. For instance, for 442 of 539 targets, an average coverage$_{RD\geq 1}$ of >80 % was achieved across all genotypes (Fig. 9.2). In comparison, for only 300 of the 539 target regions an average coverage$_{RD\geq 8}$ of >60 % of the target region was achieved across all genotypes (Fig. 9.2). These observations are in line with the non-linear relationship between RD and coverage (Fig. 9.1), and the observation that in our non-normalized pools more than half of the samples are in the non-saturating sequencing depth range.

9.4 Reproducibility in SNP Calling

Low RD may lead to a low number of SNPs passing the SNP-calling filters, and results in missing or inaccurate genotype calling data. We therefore estimated the reproducibility and accuracy of SNP calling as a function of RD. Our experimental set-up with two independent sequencing runs of a non-normalized pool of 96 samples is ideal for this analysis. On the one hand, the non-normalized pooling provides a range of sequencing depths per sample. On the other hand, because the NGS library was split after the pooling and probe capture step, approximately the same number of reads per sample was obtained in the two independent sequencing runs (inset in Fig. 9.3). We first identified a set of SNPs per sample using data of the two independent sequencing runs separately. We then compared the two SNP sets per sample using identity-by-state estimates to determine the reproducibility of SNP calling, and plotted these as function of the average RD per sample.

We identified a total of 98,047 SNP positions across the 192 (96 × 2) samples. Figure 9.3 shows the non-linear relationship between the RD and the percentage identity-by-state of technical replicates. A minimal RD of about seven is required per sample to reach >96 % identity-by-state, and a plateau is reached at about 98 % identity-by-state. Taken together, we show that SNP calling is reproducible and that targeted resequencing can effectively be used to genotype individuals. However, a small portion (2 %) of inaccuracy in SNP calling is inherent to experimental variation of NGS sequencing and may be caused by low RD, read errors, sub-optimal mapping of reads, and/or SNP mis-calling. Part of the dissimilarity between rep-

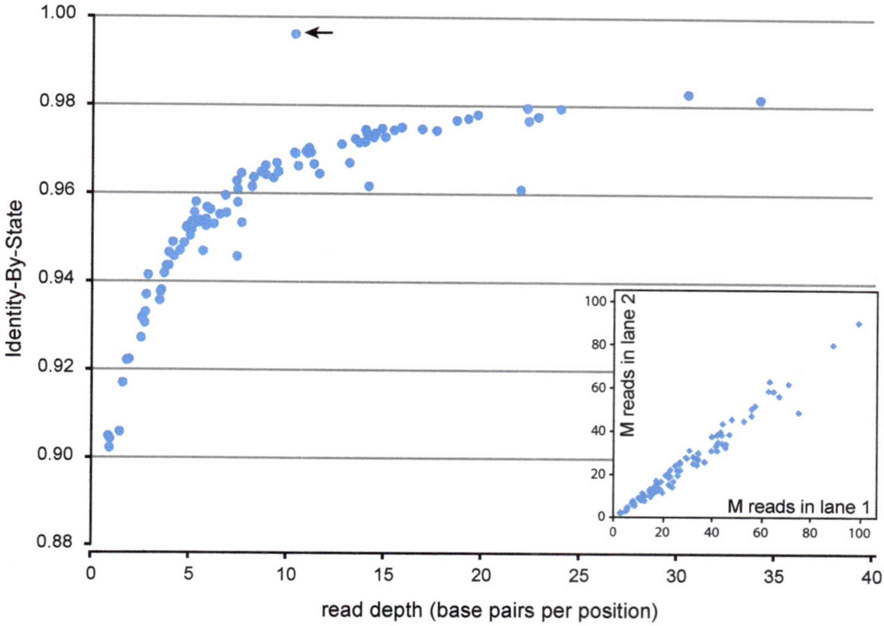

Fig. 9.3 Relationship between read depth (RD) and identity-by-state between two replicate SNP sets per sample. RD is calculated as the average of the two replicate sequencing lanes per sample. Identity-by-state was determined by comparing two SNP sets per sample obtained from two independent sequencing runs. The arrow indicates the reference genotype. Inset shows the reproducibility of the number of reads per genotype (in millions) in the two independent sequencing runs

licates is probably caused by the larger 'ambiguity' in SNP genotype calling for heterozygous positions, as suggested by the positioning of the highly homozygous reference genotype used for genome sequencing in Fig. 9.3.

9.5 Variant Calling in the Combined Dataset

Next, read mapping data from the two technical replicates were combined, rendering a total of 96,593 SNP positions and 20,696 InDel positions across the 96 samples. A certain RD is required for high-quality genotype calling for each position in each sample. As the range of sequencing depths varies across all samples, there is not always sufficient data to call the genotype for all of the 96,593 SNP positions and 20,696 InDel positions. Therefore, we investigated the relationship between the RD per sample and the number of positions with sufficient high-quality genotype data for that sample (Fig. 9.4). For example, in samples with the lowest RD (ca. 2 reads per position), for only about 65 k SNP positions and 13 k InDel positions a genotype can be called (red squares in Fig. 9.4). VCFtools provides the option to filter SNP or InDel genotype calls per sample, based on the so-called GQ score, which is, among others, dependent on the RD and mapping quality per position per

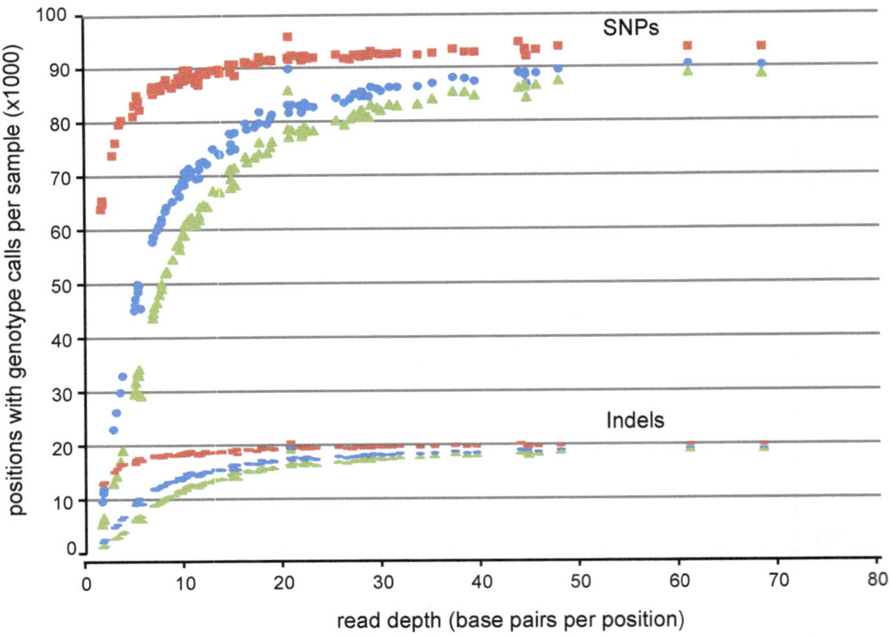

Fig. 9.4 Number of positions with genotype calls per sample (SNPs and InDels with quality scores above 30) versus read depth (RD) (calculated as the number of on-target unique reads per position and averaged over all target regions per sample). Different colors represent different cut-offs at which a particular position was retained for a particular sample (*red*: all variants retained, *blue*: variants with genotype quality (GQ) score >20, *green*: variants with GQ score >30)

genotype. The genotype calling saturation curve in Fig. 9.4 illustrates to what extent filtering for high-quality genotype calls (GQ>20 or GQ>30) reduces the total number of positions per genotype, across a range of sequencing depths. These curves show that (out of a total of 96 k (SNPs) or 20 k (InDel) polymorphic positions in the complete data set) for a given sample with median RD of 10, one can expect about 70 k SNP positions and 15 k InDel positions to be called with a GQ>20. The other positions will be marked as 'missing data' for that sample.

These 96,593 SNPs were used to reveal patterns of genetic similarity among all 96 samples using SNPhylo (Fig. 9.5). Three major groups were detected corresponding to the geographical origins of the genotypes: the Nord-Pas-de-Calais region, the Pyrenees, and the Alps. Genotypes from Bretagne were spread across the phylogenetic tree.

The inbred genotype used for genome sequencing and as reference for read mapping clustered relatively distant from the rest. This suggests that the reference genome sequence is substantially different from the other samples, which may affect read mapping. However, detailed analysis of the 96,593 SNP set revealed that only 236 SNPs were unique for the reference genotype and were monomorphic (for the alternative allele) across all French genotypes. Conversely, 96,357 SNPs were polymorphic across the French accessions after elimination of the reference genotype from the data set. We therefore conclude that this apparent genetic dissimilarity is caused by the

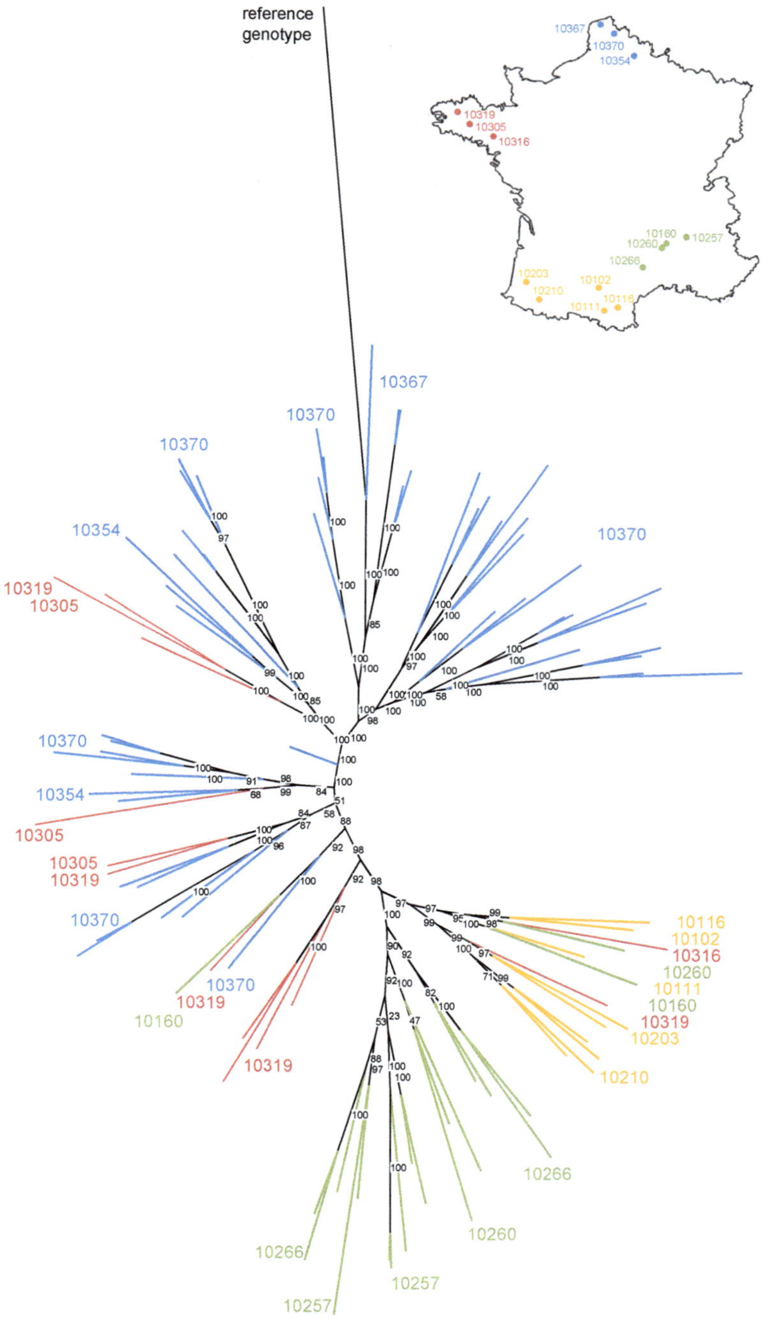

Fig. 9.5 Phylogenetic relationships among the set of 96 genotypes based on 96,593 SNPs (maximum likelihood using 100 bootstraps). The different colors represent different geographical origins in France as shown in the inset figure (*blue*: Nord-Pas-de-Calais, *red*: Bretagne, *green*: Alps, *yellow*: Pyrenees)

high degree of homozygosity in the inbred reference genotype (inbreeding coefficient F=0.94), and the high degree of heterozygosity (F<0.5) of all other samples.

The median polymorphism density across 539 target regions was about 3.56 SNPs/100 bp and 0.72 InDels/100 bp, which is in line with previous SNP density estimates based on large scale transcriptome data (Ruttink et al. 2013). Although the values varied within gene families and pathways or physiological functions, no clear differences were found between pathways (Fig. 9.6). No correlation was detected between the number of SNPs and the number of InDels in the target regions (Fig. 9.6). This high polymorphism density stresses once more the relevance of resequencing approaches for genotyping of highly diverse genotype panels as the one investigated here.

9.6 Linkage Disequilibrium (LD)

The LD was estimated for each target region separately, and expressed as the average distance in base pairs between a SNP pair for which an r^2-value of 0.2 was reached. The calculations were performed for two subpopulations, to accommodate for the large number of plants (44) of accession 10370 genotyped in this study. Furthermore, this allowed to compare LD decay values within a group of related plants (subpopulation 2) to that within a group of less related plants (subpopulation 1). LD estimates were quite similar for both subpopulations (Table 9.1). However, for subpopulation 2, the LD decays over a slightly but significantly longer distance ($p<0.05$). This is not surprising if we consider that the average kinship is expected to be larger in subpopulation 2 than in subpopulation 1. However, the differences in LD between the subpopulations are not large, indicating that LD patterns over short distances are similar in a collection of plants from a single accession (representing one sampling location) and a collection of plants of diverse origins (17 sampling locations across France). As expected, at higher MAF, LD decays over longer distances. The lower the number of rare variants considered, the higher the average correlation between SNP pairs, and therefore, the longer the distance over which LD spans.

In both subpopulations, r^2 values dropped below 0.2 at distances below 1000 bp for most target regions (Fig. 9.7). This confirms LD decay over short distances

Table 9.1 LD decay estimations (in bp) for two subpopulations and for the total set of genotypes (for further details see main text). Intra-target region average r^2 values were estimated at different MAF. The length (in bp) at which the fitted r^2 curve took a value of 0.2 for each target region was used to estimate these average values

	MAF=0	MAF=0.05	MAF=0.10
Subpopulation 1	834	933	1025
Subpopulation 2	897	984	1172
All	812	920	1128

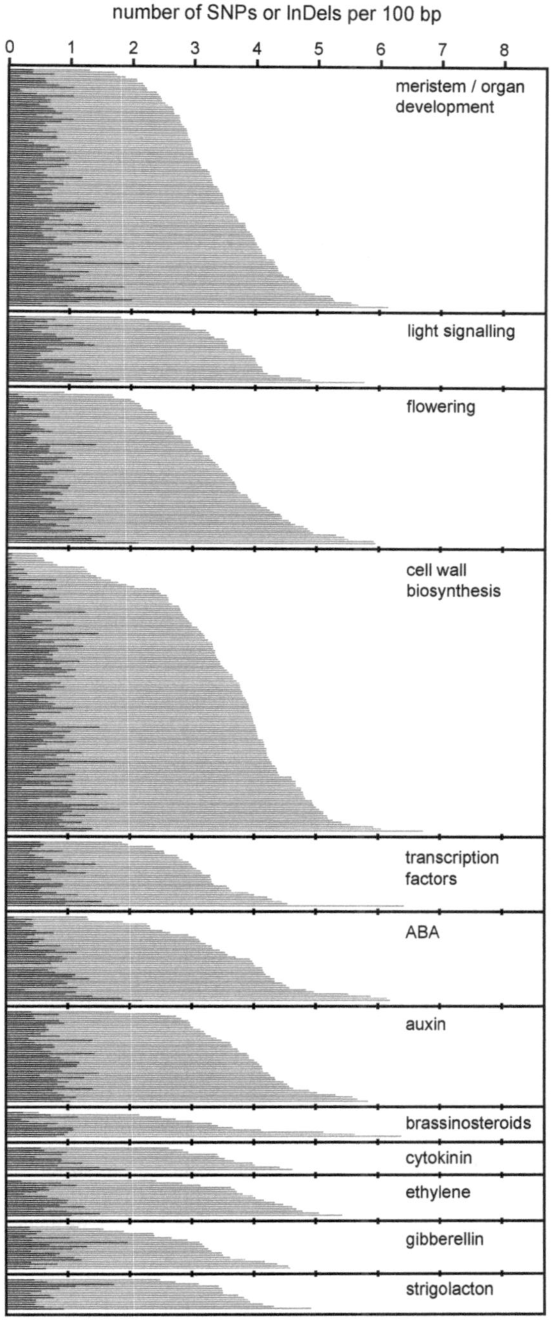

Fig. 9.6 Number of SNPs (*light grey bars*) and InDels (*dark grey bars*) per gene identified in this study. The genes have been classified according to the physiological process/pathway in which they are involved, and ordered according to the number of SNPs detected

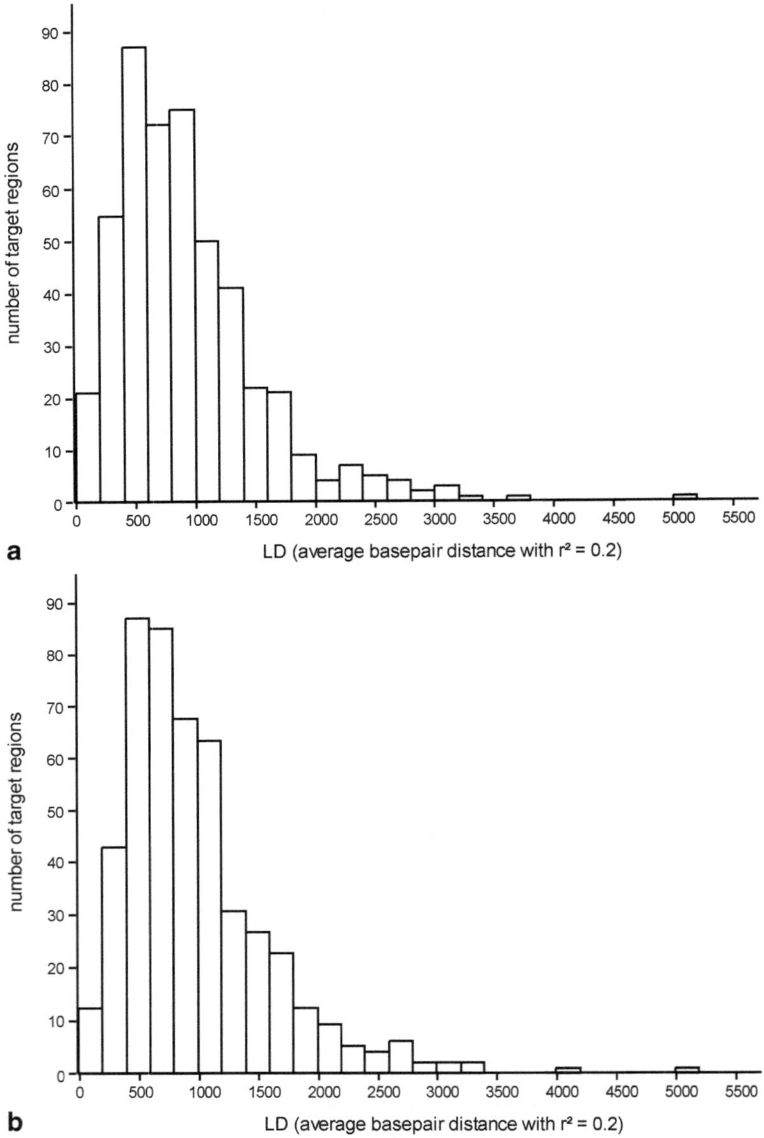

Fig. 9.7 Overview of LD patterns across different genic regions in the *L. perenne* genome at MAF=0.05. A LD decay plot was constructed for each target region and the distance (in bp) at which the fitted r^2 curve took a value of 0.2 was identified for each of them. A. Subpopulation 1; B. Subpopulation 2

in perennial ryegrass as previously reported in various populations (ranging from 300 bp to 2000 bp according to Skøt et al. 2007; Yu et al. 2013; Xing et al. 2007; Brazauskas et al. 2011; Fiil et al. 2011).

9.7 Conclusions and Future Perspectives

The results presented here demonstrate that the probe set used in this study is useful for probe capture enrichment of genomic shotgun libraries followed by NGS. Furthermore, our in-house developed custom procedure for identification and elimination of probes with high sequence similarity to non-targets, repetitive sequences, or mitochondrial or chloroplast sequences allows recovery of a large percentage (>95 %) of the intended target regions, given sufficient sequence depth. It further shows that the current state of the draft genome sequence provides a rich resource for comprehensive gene discovery and gene family analysis.

The approach followed is useful to create a comprehensive inventory of the sequence diversity in >500 *L. perenne* genes. This information could be used for AM, but it is also a powerful tool to de novo discover naturally occurring sequence variants that potentially disturb gene function. Genotypes carrying such defective alleles may be useful in reverse genetic approaches to study gene function, or for breeding as demonstrated in poplar (Vanholme et al. 2013).

Careful experimental design for pooling is required to achieve sufficient sequencing depth per genotype and sufficient RD per position. When performing targeted resequencing a balance must be found between the number of samples to pool in a sequencing run, the total length of target sequence and the minimal RD required for reliable polymorphism calling. Clearly, increasing the number of genotypes or the length of target sequence negatively affects the sequencing depth per genotype. Normalization of samples in pooled libraries, reduction of the number of samples per pool, reduction of the intended target length, sequencing multiple lanes per pool, or optimizing the post-hybridization PCR-amplification conditions to reduce the number of PCR-duplicates (Rohland and Reich 2012; Bodi et al. 2013) may all contribute to saturating target length coverage.

However, the relationship between RD and target coverage is non-linear. Predicting the minimal required sequencing depth per sample is not trivial, but future experiments can be based on the coverage and genotype calling saturation curves presented here. These data illustrate the possible trade-off scenarios. In planning an experiment for a particular population one can choose between aiming to sequence 2.7 Mbp target region in a pool of 96 genotypes at the price of partial coverage (e.g. 60 % of the intended target length at RD$\geq$8), or aiming for more complete (80–90 %) coverage per target region at the price of sequencing for instance four times less candidate genes (e.g. only 700 k bp target region), or a four times smaller pool (24 genotypes) per sequencing lane.

We have used the relationships between different parameters presented in this paper as guideline for a larger targeted resequencing experiment in which about 600 extra genotypes are screened. These genotypes have been characterized for plant developmental traits and cell wall characteristics during two seasons. These results are currently being analysed and will be reported elsewhere.

Acknowledgements The data analysis was carried out using the STEVIN Supercomputer Infrastructure at Ghent University, funded by Ghent University, the Flemish Supercomputer Center

(VSC), the Hercules Foundation, and the Flemish Government—department EWI. This research was supported by the Agency for the promotion of Innovation by Science and Technology (IWT) in Flanders, Belgium (project LO-080510), and by Aarhus University and the Godfred Birkedal Hartmann foundation in Denmark.

References

Auzanneau J, Huyghe C, Julier B, Barre P (2007) Linkage disequilibrium in synthetic varieties of perennial ryegrass. Theor Appl Genet 115:837–847

Bertone P, Trifonov V, Rozowsky JS, Schubert F, Emanuelsson O, Karro J, Kao M-Y, Snyder M, Gerstein M (2006) Design optimization methods for genomic DNA tiling arrays. Genome 16:271–281

Bodi K, Perera AG, Adams PS, Bintzler D, Dewar K, Grove DS, Kieleczawa J, Lyons RH, Neubert TA, Noll AC, Singh S, Steen R, Zianni M (2013) Comparison of custom target enrichment methods. J Biomol Tech 24(2):73–86

Bolger AM, Loshse M, Usade B (2014) Trimmomatic: a flexible trimmer for illumina sequence data. Bioinformatics 30(15):2114–2120. doi:10.1093/bioinformatics/btu170

Brazauskas G, Lenk I, Pedersen MG, Studer B, Lübberstedt T (2011) Genetic variation, population structure, and linkage disequilibrium in European elite germplasm of perennial ryegrass. Plant Sci 181:412–420

Byrne S, Armstead I, Swain SC, Panitz F, Bendixen C, Hedegaard J, Nagy I, Farrell JD, Studer B, Caccamo M, Asp T (2014) A Draft Sequence Assembly of the Perennial Ryegrass (*Lolium perenne*) Genome. Abstract Plant and Animal Genome Meeting XXI. https://pag.confex.com/pag/xxii/webprogram/Paper10555.html

Danecek P, Auton A, Abecasis G, Albers CA, Banks E, DePristo MA, Handsaker RE, Lunter G, Marth GT, Sherry ST, McVean G, Durbin R (2011) 1000 genomes project analysis group the variant call format and VCFtools. Bioinformatics 27(15):156–2158

Davey JW, Hohenlohe PA, Etter PD, Boone JQ, Catchen JM (2011) Genome-wide genetic marker discovery and genotyping using next-generation sequencing. Nat Rev Genet 12:499–510

Fiil A, Lenk I, Petersen K, Jensen CS, Nielsen KK, Schejbel B, Andersen JR, Lubberstedt T (2011) Nucleotide diversity and linkage disequilibrium of nine genes with putative effects on flowering time in perennial ryegrass (*Lolium perenne* L.). Plant Sci 180:228–237

Lee T-H, Guo H, Wang X, Kim C, Paterson AH (2014) SNPhylo a pipeline to construct a phylogenetic tree from huge SNP data. BMC Genome 15:162

Letunic I, Bork P (2011) Interactive tree of life v2: online annotation and display of phylogenetic trees made easy. Nucleic Acids Res 39:W475–478. doi:10.1093/nar/gkr201

Li H, Durbin R (2010) Fast and accurate long-read alignment with Burrows—Wheeler transform. Bioinfomatics 26(5):589–595

McKenna A, Hanna M, Banks E, Sivachenko A, Cibulskis K, Kernytsky A, Garimella K, Altshuler D, Gabriel S, Daly M, DePristo MA (2010) The genome analysis toolkit a MapReduce framework for analyzing next-generation DNA sequencing data. Genome Res 20:1297–1303

Rohland N, Reich D (2012) Cost-effective, high-throughput DNA sequencing libraries for multiplexed target capture. Genome Res 22:939–946

Ruttink R, Sterck L, Rohde A, Bendixen C, Rouze P, Asp T, Van de Peer Y, Roldan-Ruiz I (2013) Orthology guided assembly in highly heterozygous crops: creating a reference transcriptome to uncover genetic diversity in *Lolium perenne*. Plant Biotechnol J 11(5):605–617

Skøt L, Humphreys J, Humphreys MO, Thorogood D, Gallagher J, Sanderson R, Armstead IP, Thomas ID (2007) Association of candidate genes with flowering time and water-soluble carbohydrate content in (L.). Genetics 177:535–547

Skøt L, Sanderson R, Thomas A, Skot K, Thorogood D, Latypova G, Asp T, Armstead I (2011) Allelic variation in the perennial ryegrass FLOWERING LOCUS T gene is associated with changes in flowering time across a range of populations. Plant Physiol 155:1013–1022

Uitdewilligen JGAML, Wolters A-M A, D'hoop BB, Borm TJA, Visser RGF, van Eck HJ (2013) A next-generation sequencing method for genotyping- by-sequencing of highly heterozygous autotetraploid potato. PlosONE 8(5):e62355

Van Bel M, Proost S, Wischnitzki E, Movahedi S, Scheerlinck C, Van de Peer Y, Vandepoele K (2012) Dissecting plant genomes with the PLAZA comparative genomics platform. Plant Physiol 158:590–600

Vanholme B, Cesarino I, Goeminne G, Kim H, Marroni F, Van Acker R, Vanholme R, Morreel K, Ivens B, Pinosio S, Morgante M, Ralph J, Bastien C, Boerjan W (2013) Breeding with rare defective alleles (BRDA): a natural *Populus nigra* HCT mutant with modified lignin as a case study. New Phytol 198:765–776

Wendl MC (2006) A general coverage theory for shotgun DNA sequencing. J Comput Biol 13(6):1177–1196

Xing Y, Frei U, Schejbel B, Asp T, Lübberstedt T (2007) Nucleotide diversity and linkage disequilibrium in 11 expressed resistance candidate genes in *Lolium perenne*. BMC Plant Biol 7:43

Yu X, Bai G, Liu S, Luo N, Wang Y, Richmond DS, Pijut PM, Jackson SA, Yu J, Jiang Y (2013) Association of candidate genes with drought tolerance traits in diverse perennial ryegrass accessions. J Exp Bot 64(6):1537–1551

Zheng X, Levine D, Shen J, Gogarten SM, Laurie C, Weir BS (2012) A high-performance computing toolset for relatedness and principal component analysis of SNP data. Bioinformatics 28(24):3326–3328

Zhu C, Gore M, Buckler ES, Yu J (2008) Status and prospects of association mapping in plants. Plant Genome 1:5–20

Chapter 10
Advances in Forage and Turf Biotechnology in China

Shuangyan Chen, Yanmin Wu, Tao Wang, Suomin Wang, Yongzhen Pang, Xinquan Zhang, Meiliang Zhou, Jiangli Dong, Aike Bao, Linkai Huang and Gongshe Liu

10.1 Introduction

Grasslands are among the largest ecosystems on the world representing almost 40 % of the terrestrial area. There are 393 million ha of natural grasslands in China, accounting for 41.7 % of the total terrestrial area. China has abundant forage grasses. The forage species are critical to livestock industries and have many advantages such as high environmental adaptation to various stresses of drought, salt, and cold and play important roles in preventing soil erosion and grassland degeneration and desertification. Given species diversity and the advantages of forage in China, research and utilization of forage species are still limited because of the less research funds of forage compared to crop species, the polyploid and complexity genome,

G. Liu (✉) · S. Chen · Y. Pang
Key Laboratory of Plant Resources, Institute of Botany, Chinese Academy of Sciences, Beijing, P. R. China
e-mail: liugs@ibcas.ac.cn

Y. Wu · M. Zhou
Biotechnology Research Institute, Chinese Academy of Agricultural Sciences, 100081 Beijing, P. R. China

T. Wang · J. Dong
State Key Laboratory of Agrobiotechnology, College of Biological Sciences, China Agricultural University, 100193 Beijing, P. R. China

S. Wang · A. Bao
State Key Laboratory of Grassland Agro-ecosystems, College of Pastoral Agriculture Science and Technology, Lanzhou University, 730020 Lanzhou, P. R. China

X. Zhang · L. Huang
Department of Grassland Science, Animal Science and Technology College, Sichuan Agricultural University, Ya'an, P. R. China

H. Budak, G. Spangenberg (eds.), *Molecular Breeding of Forage and Turf*,
DOI 10.1007/978-3-319-08714-6_10

less genomic information, the difficulty of genetic transformation, and the feature of perennial.

The advent of high-throughput next-generation sequencing (NGS) technologies, such as Roche/454, Illumina/Solexa, and ABI/SOLiD, has made it possible to generate genome resources at a large scale and relatively low cost. However, whole-genome sequencing is currently expensive and impractical for most of forage. At present, transcriptome sequencing is an efficient strategy to explore the forage gene resources. The new sequencing technology also accelerates forage molecular markers mining and the molecular breeding. A lot of biotechnology methods, which have been successfully applied in model and crop species, have been applied for forage improvement.

So far, many achievements have been made in China in forage biotechnology including transcriptome-based gene mining, molecular marker technologies, and genetic engineering. This review gives an introduction of the most recent developments in biotechnology and their applications in genetic improvement of forage grasses in China.

10.2 Transcriptome-Based Gene Mining

The application of NGS to transcriptomics, commonly called RNA-seq, allows the nearly complete characterization of transcriptomic events occurring in a specific tissue. It has proven particularly useful in non-model species. The DNA sequence of highly heterozygous, large, and complex genomes of allogamous forage and turf grass species is difficult to access, even with NGS technology. An efficient strategy for complexity reduction is transcriptome sequencing since the focus of sequencing is restricted to the coding regions, rather than the entire genome.

Sheepgrass (*Leymus chinensis* (Trin.) Tzvel.) is an important perennial forage grass across the Eurasian Steppe, and widely distributed on the Northeast Plain, the Northern Plain, and the Inner Mongolian Plateau of China. It is known for its adaptability to various environmental conditions and plays important roles in soil and water conservation, ecological construction and support of livestock farming, especially in northern China. The transcriptome of sheepgrass was sequenced using both Roche 454 pyrosequencing technology and Illumina/Solexa platform (Chen et al. 2013; Chen et al. 2014). A total of 87,214 unigenes were assembled, including 32,416 contigs and 54,798 singletons, and 54,584 (62.6%) unigenes were annotated. There were 11,675 core Poaceae-specific and 12,811 putative sheepgrass-specific unigenes identified by BLAST searches against all plant genome and transcriptome databases. A total of 2979 freezing-responsive unigenes were identified by comparative transcriptome analysis of differentially expressed genes (DEGs; Chen et al. 2013). In total, 1836 and 3238 genes were detected with significant differential expression levels after wounding and defoliation in 1 day, respectively, based on a comparative transcriptome analysis of the Illumina/Solexa sequencing data. According to the results of GO, KOG and pathway-based enrichment analyses of the DEGs, we demonstrated that both wounding and defoliation activated the systemic synthesis of jasmonate (JA). However, defoliation specifically reduced the expression levels of ribosomal protein genes, cell division or cell expansion-related genes, and lignin biosynthesis genes and may have negatively affected plant growth. Further experimental data revealed that the regrowth of elongating leaves was significantly retarded after defoliation at 6 h through the following 7 days of measurement, suggesting that the gene expression pattern and phenotype are consistent (Chen et al. 2014). The identified stress-related genes will help us to decipher the genetic basis of the environmental and ecological adaptations of this species and will be used for forage genetic improvement. A lot of stress-related genes, such as *LcDREB3a*, LcDREB2, *LcSAMDC2*, *LcSAIN1*, *LcSAIN2*, *LcMYB1*, *LcSUT1*, had been cloned and identified from this sheepgrass dataset (Peng et al. 2011; Peng et al. 2013; Li et al. 2013a; Li et al. 2013b; Cheng et al. 2013; Su et al. 2013).

Medicago truncatula is a model legume whose genome has been sequenced. Li et al. (2009) developed a database called MtED that contains gene expression profiles of the roots of *M. truncatula* based on time-course salt stress experiments using the Affymetrix Medicago GeneChip. They identified genes of *MtCBF4* and *MtCAS31* from the microarray expression profiles data, and showed that the two genes play important role in abiotic stress responses (Li et al. 2011; Xie et al. 2012).

Liu et al. (2013a) explored the floral transcriptome of *Vicia sativa*, and showed that the vetch flowers conform to a strict ABC model. They also analyzed the evolution and expression of the TCP gene family in vetch at a whole-genome level, and several unigenes specific to three different vetch petals, which might offer some clues toward elucidating the molecular mechanisms underlying floral zygomorphy.

10.3 Molecular Marker Technologies

Molecular markers, as particular segments of DNA that represent different genome levels, play an essential role today in all aspects of plant breeding, and have been widely used in genetic map, targeting genes, and marker-assisted selection (MAS). Comparing with conventional phenotype, molecular markers had numerous advantages as they are easily detectable and stable in plant tissues regardless of the influences caused by environment, epstatic or pleiotropic effects. At present in China, molecular marker technologies are mainly employed in evaluation of genetic diversity, evolution and relationship in main forage and turf grass including *Leymus chinensis*, orchardgrass, *Hemarthria*, alfalfa (*Medicago sativa* L.), *Lolium*, and *Cynodon dactylon*. Among numerous molecular markers, simple sequence repeats (SSR), sequence-related amplified polymorphism (SRAP), inter simple sequence repeat (ISSR), and random amplified polymorphic DNA (RAPD) have been a hot area of research for recent years in China.

SSR markers, as an important method targeting known sites, first emerged in plants around 1990. Alfalfa is the most widely cultivated forage legume and one of the most economically valuable crops in the world. Liu et al. (2013b) identified

1649 potential EST-SSRs from Illumina transcriptome sequencing alfalfa dataset. A total of 100 pairs of PCR primers were randomly selected to develop EST-SSR markers from genomic DNA. Of these primer pairs, 82 were able to amplify sequences and 27 primer pairs were detected significant amounts of polymorphism among 10 alfalfa accessions. Chen et al. (2013) identified 3818 EST-SSRs in 3597 unigenes from the 454 transcriptome dataset of sheepgrass, and some SSRs contained unigenes that were also candidates for freezing-response genes. Jiang et al. (2013) used SSR markers to construct DNA fingerprinting of orchardgrass cultivars and found 11 bands which can be used for the development of DNA fingerprints in 32 orchardgrass cultivars. Another study focused on the estimation of genetic variability in orchardgrass using cereal EST-SSRs and orchardgrass SSR markers. In total, 190 polymorphic bands were detected and the average polymorphic rate (P) for the species was 84.63 %, indicating high degree of genetic diversity. Through the analysis in distinct geographical divergence of orchardgrass, this study also indicated that northern Africa, Europe, and temperate Asia might be the diversity differentiation center of orchardgrass (Xie et al. 2010). Further, SSR markers have proven to be highly polymorphic, however, as these require sequence data for each locus, causing slow development in new forage and turf grass species.

In assessment of genetic diversity, SRAP markers have been determined as the useful molecular marker system owe to their reliability, simplicity, moderate throughput ratio, and easy sequencing of selected bands. In south China, *Hemarthria compressa* is one of the most important and widely utilized forage crops due to high adaptability to hot and humid conditions and high forage yield. Twelve populations of *H. compressa* in south China using SRAP markers, and high-genetic diversity was found in these materials (Huang et al. 2012). SRAP molecular markers were also used in *C. dactylon,* which is a main warm season turf grass, for the identification of nine *C. dactylon* varieties (Liu et al. 2012). Furthermore, SRAP target coding sequences in the genome and bring about numerous co-dominant markers, which are helpful for the development in map construction. A diploid ($2n = 2x = 14$) orchardgrass genetic linkage map was constructed using SRAP and SSR markers. The paternal genetic map contains 90 markers (57 SSRs and 33 SRAPs) over 9 linkage groups (LGs), while the maternal genetic map is composed of 87 markers (54 SSRs and 33 SRAPs) assembled over 10 LGs, which provide a guidance in MAS and mapping quantitative trait locus (QTLs) of agronomic traits in orchardgrass (Xie et al. 2011).

RAPD is a rapid and inexpensive method which does not need any information about the genome of the plant, and it is always supposed to the investigation of genetic variability in numerous plants. RAPD markers were first used for the researches on genetic diversity and germplasm identification in alfalfa (Liu et al. 2010), which is widely cultivated around the world and has been one of the oldest cultivated as well as wild crops in China. Jiang et al. (2011) analyzed the genetic structure and diversity of 25 alfalfa varieties using RAPD technology, and found that single-plant DNA sample with polymorphism loci percentages (81.52 %) was a better method than mixed DNA sample polymorphism loci percentages (61.65 %), to reveal the level of genetic variation within and between alfalfa varieties. Ma et al. (2013a) examined genetic relationships among 18 accessions which represent all

Lolium species using RAPD markers. A total of 367 RAPD fragments were detected, of which 95.9 % were polymorphic across all the *Lolium* accessions, indicating high-genetic diversity in these accessions. Another valuable finding in this study is that *L. canariense* formed a distinct subclade, which showed higher genetic separation from other allogamous species.

10.4 Advances of Genetic Engineering

10.4.1 Quality Improvement

Advances in forage quality improvement in China among forage grass community mainly include but not limited to the following traits: lignin, essential amino acids (EAAs), soluble carbohydrates, condensed tannins, sapnins, and vitamin E. Biotechnology techniques have been explored to introduce genes controlling the above-mentioned traits to improve forage quality, by repressing or activating the expression levels of target genes.

Lignin (polymers of monolignols), one of the major components of cell wall, inhibits forage digestibility by ruminant animals. The biosynthetic pathway of lignin is well-documented, which makes it possible to control the content and composition of lignin in forage crops, by manipulating the expression levels of key genes in the pathway. Cinnamoyl-CoA reductase (CCR), cinnamyl alcohol dehydrogenase (CAD), and 4-coumarate: CoA ligase (4CL) are the three key enzymes involved in the biosynthesis of monolignols. By using RNAi technique, Hu and coworkers found that the lignin content was reduced significantly in ryegrass (*Lolium perenne*) when CCR and CAD genes were silenced, respectively (Hu et al. 2013). Similarly, 4CL and CAD genes were cloned from *Pennisetum purpureum* cv. Huann., and they were trying to down-regulate the lignin level by silencing these target genes in *P. purpureum* (Huo et al. 2012; Tang et al. 2014).

Alfalfa is lack of a biosynthetic pathway for methionine, cysteine, lysine, and other EAAs in vivo, which animals must obtain these EAAs from their diet. And the sulfur amino acid (SAA), mainly methionine and cysteine, are particularly important EAAs, could improve the quality and production of milk and wool and are considered as an index of nutritional value of alfalfa. Aspartate kinase (AK) catalyzes the phosphorylation of aspartate to Asp-phosphate, the first step in the aspartate family biosynthesis pathway, and adenosine 5′-phosphosulphate reductase (APR) catalyzes the conversion of activated sulfate to sulfite, providing reduced sulfur for the synthesis of cysteine, methionine, and other essential metabolites and secondary compounds. So AK and APR are the key enzymes of aspartate family biosynthesis pathway and sulfate assimilation pathway respectively. A transformed vector with AK and APR was constructed and then transformed cotyledon of alfalfa mediated by *Agrobacterium tumefaciens*. Transgenic plants had no significant changes in phenotype and agricultural properties. Amino acids analysis revealed that Cys increased by 28.6%, Met increased by 58.8%, Asp increased by 30.7–40.4%, and Lys increased by 38.7–51.6%, but Pro decreased by 35.6–40.0%, Ser decreased by about 10%, and OAS increased by 21.0–47.5% in the T0 transgenic plants. There is a similar result in T1 transgenic progeny; and, changes of amino acids contents connected with TCA circle through oxaloacetate and α–ketoglutarate, the increase of Asp family accompanies with the declined of Glu family. In addition, an increase in cysteine was likely at the expense of serine, the carbon skeleton donor of cysteine (Tong et al. 2014). This approach provides a new method for increasing SAAs and allows for the development of new genetically modified crops with enhanced nutritional value.

Fructan is one of the major soluble carbohydrates present in forage grasses, the increase of which could enhance grass absorption and digestibility in ruminant animals. A fructan:fructan 1-fructosyltransferase gene (*Ac1-FFT*) from *Agropyron cristatum* was transformed into forage grass *L. perenne* by agrobacterium-mediated method (Zhang et al. 2011). The results showed that both water soluble carbohydrates and fructan contents were significantly higher in Ac1-FFT-expressing transgenic *L. perenne* plants than in the non-transformed ones. In addition, the drought tolerance of transgenic *L. perenne* plants was enhanced by over-expressing *Ac1-FFT* gene (Zhang et al. 2011). This study demonstrated that this gene could be utilized not only for nutrition improvement but also for stress tolerance in forage grasses.

Condensed tannins are oligomeric or polymeric compounds of flavan-3-ols derived from the general flavonoid biosynthetic pathway. A moderate condense tannin level is an important quality trait in forage crops, because condensed tannins can protect ruminant animals from lethal pasture bloat by binding proteins and thereby slowing down their fermentation in the rumen. Flavanone-3-hydroxylase gene, which encodes an enzyme involved in the early condensed tannin pathway, was characterized from *M. truncatula* (Shen et al. 2010). Dihydroflavonol-4-reductase (DFR) gene, one of the key genes in the condensed tannin pathway, was also characterized from *Medicago sativa* (Dong et al. 2012). Over-expression of DFR in *M. truncatula* could enhance the accumulation of anthocyandin, the precursor for condensed tannin biosynthesis (unpublished data). Pang et al. (2008) discovered a new UDP-glucosyltransferase that is specific for flavan-3-ol glycosylation from *M. truncatula*. They further revealed that the glycosylation of flavan-3-ol monomer could prevent the oxidation of condensed tannins. More importantly, the condensed tannin content was increased if the oxidation process of condensed tannins could be inhibited in *Arabidopsis* (Pang et al. 2013), so this strategy could be applied in *M. sativa* for condensed tannin improvement.

In addition to the structural gene involved in the condensed tannin pathway, key transcription factors were also the targets for genetic engineering of condensed tannins in forage. Although WD repeat type transcription factor WD40-1 is necessary from condensed tannin biosynthesis in *M. truncatula*, but over-expression of *MtWD40-1* failed to increase condensed tannin in alfalfa foliage (Pang et al. 2009). This finding, together with other related investigation, demonstrated that a single component of a regulatory complex is not sufficient for condensed tannins accumulation by genetic engineering.

Saponins, glucosides of triterpenoid, are the main antinutritional components present in protein rich alfalfa plants. Isoprenoid is the precursor for the biosynthesis of saponins, the biosynthetic pathway of which has been characterized in some model plants. Three isoprenyl diphosphate synthases genes (FPPS, GPPS, and GGPPS) were isolated and characterized from alfalfa, and *MsFPPS* was confirmed to be able to increase squalene level in transgenic tobacco (Sun et al. 2013). The availability of these three genes makes it possible to further modify sapnin level in alfalfa in combination with other later pathway genes.

Tocopherols, also known as vitamin E, are one of the major nutritional factors in forage grasses. Since the biosynthetic pathway of tocopherols in model plant *Arabidopsis* has been fully characterized, vitamin E content and composition changes could be achieved more easily in forage grasses. A key gene γ-tocopherol methyltransferase (TMT) in the last step of vitamin E biosynthesis was characterized from *M. sativa* (Jia et al. 2012). The over-expression of *MsTMT* gene increased the level of α-tocopherol, the most active tocopherol component, but the total tocopherol content was unchanged (unpublished data). In addition to *MsTMT* gene, the ongoing characterization and over-expression of tocopherol cyclase, hydroxyphenylpyruvate dioxygenase and homogentisate phytyltransferase genes in alfalfa, will provide new germplasm for genetic breeding of high value forage grasses with increased vitamin E level.

10.4.2 Abiotic Stress Tolerance

Halophytes are the plant species that can survive and reproduce in the growing conditions with above 200 mM NaCl and comprise approximately 1 % in the plant kingdom. Some halophytes, such as *Suaeda maritima*, require saline environments for optimal growth, while others, such as *Puccinellia tenuiflora*, do not require saline conditions for optimal growth but can tolerate high concentrations of salts. Therefore, various types of halophyte may utilize specific mechanisms to adapt to salinity.

P. tenuiflora is a halophytic monocotyledonous species found in saline marshes of north China and is used as forage as well as for soil improvement. Compared to wheat, *P. tenuiflora* exhibits stronger selective absorption and transport for K^+ over Na^+. Guo et al. (2012a) cloned the *PtSOS1* gene encoding plasma Na^+/H^+ antiporter (SOS1) from *P. tenuiflora* and found that the expression levels of *PtSOS1* in roots increased significantly with the increase of external NaCl (25–150 mM), accompanied by an increase of selective transport (ST) capacity for K^+ over Na^+ by roots. Transcription levels of *PtSOS1* in roots and ST values increased under 0.1–1 mM KCl, then declined sharply under 5–10 mM KCl. Under 150 mM NaCl, *PtSOS1* expression levels in roots and ST values at 0.1 mM KCl was significantly lower than that at 5 mM KCl with the prolonging of treatment time. A significant positive correlation was found between root *PtSOS1* expression levels and ST values under various concentrations of NaCl, KCl or 150 mM NaCl plus 0.1 or 5 mM KCl treatments. These results suggested a role of *PtSOS1* in the plasma membrane of the xylem parenchyma cells in controlling K^+ transport into and Na^+ retrieval from the xylem and proposed that *PtSOS1* could be a major player in salt tolerance via ST for K^+ over Na^+ in *P. tenuiflora*. They also hypothesize a function model of SOS1 in regulating K^+ and Na^+ transport system in the membrane of xylem parenchyma cells by sustaining the membrane integrity, which could reasonably explain the phenomenon of Na^+ retrieval from the xylem when plants are exposed to severe salt stress (Guo et al. 2012a).

Zygophyllum xanthoxylum, perennial fodder shrubs with high palatability and nutrient value that colonizes arid areas in China and Mongolia, are typical succulent xerophytes with excellent adaptability to adverse arid environments. These species could accumulate larger quantities of Na^+ through roots and transport Na^+ to leaves for osmotic adjustment even at low-salt soil, suggesting that Na^+ may have a positive function in *Z. xanthoxylum* response to water deficit. To investigate the molecular mechanisms how Na^+ efficiently transported and accumulated to the leaves, *ZxSOS1* and *ZxNHX* encoding plasma and tonoplast Na^+/H^+ antiporter were cloned, respectively (Wu et al. 2011; Ma et al. 2013b). The results from post-transcriptional gene silencing experiments demonstrated that *ZxSOS1* is not only essential in long-distance transport and spatial distribution of Na^+ and even K^+ but also vital for regulating K^+ and Na^+ transport system and maintaining Na^+ and K^+ homeostasis in *Z. xanthoxylum*, thereby regulating its normal growth (Ma et al. 2013b). Wu et al. (2011) found a positive correlation between up-regulation of *ZxNHX* and accumulation of Na^+ in *Z. xanthoxylum* exposed to salt or drought, suggesting that tonoplast Na^+/H^+ antiporter plays important roles in Na^+ accumulation and homeostasis of *Z. xanthoxylum* under salt and drought conditions. Bao et al. (2014) cotransferred *ZxNHX* and *ZxVP1-1* (encoding tonoplast H^+-PPase) genes from *Z. xanthoxylum* into the important legume forages *Lotus corniculatus* and alfalfa. The transgenic lines coexpressing above genes showed enhanced salt and drought tolerance, which were associated with higher shoot biomass, more cations accumulation, higher water retention, lesser cell membrane damage and higher photosynthesis capacity under salt or water-deficit condition.

10.4.3 Development of Transgenic Vaccine

Development of forage plants as host systems for production of viral antigens has been reported in the past (Wang et al. 2008; Rao et al. 2012; Guo et al. 2012b). This strategy has several advantages such as cost effectiveness, absence of processing and thermostability. In addition, the technology provides an opportunity for mass immunization as the edible parts of the forage plant can be directly fed.

Foot and mouth disease virus (FMDV) is the causative agent of an economically important disease affecting meat-producing animals. Several reports showed that

FMDV VP1 protein of single serotype has been mostly used as immunogen (Wang et al. 2008; Rao et al. 2012). *Stylosanthes* spp. is an important commercial forage legume because of its high crude protein content, thus, it could be considered as an ideal host system of animal plant-based vaccines. In China, Wang and colleagues developed an oral immunization system for foot and mouth disease with the structural protein VP1 of FMDV produced in transgenic forage plant *Stylosanthes guianensis* cv. Reyan II as feedstuff additives. Enzyme-linked immunosorbent assays (ELISA) analysis showed that the level of the expressed protein VP1 in transgenic plants was 0.1–0.5 % total soluble protein. Their results demonstrated that mice that were orally immunized with the transgenic plants as feedstuff additives developed a virus-specific immune response to the structural VP1 and intact FMDV particles (Wang et al. 2008).

Avian influenza virus caused disease not only in poultry but also in humans. The H5N1 avian influenza virus is a highly contagious and deadly pathogen in domestic fowl. Hemagglutinin (HA), the principal viral surface antigen, is the primary target for neutralizing antibodies. It is responsible for binding of the virus to host receptors, enabling entry into the host cell through endocytosis and subsequent membrane fusion. *Lotus corniculatus* L. (birdsfoot trefoil), a perennial herbaceous forage plant, it is an attractive expression system for the economic production of antigen proteins because of its excellent protein quality with a lower cellulose content and more carbohydrate than clover and alfalfa. Wu and his colleagues first reported that over-expression of the avian influenza (H5N1) HA gene in *L. corniculatus* can be detected by western blotting and ELISA. However, the maximum expression level of the HA protein is up to 0.00786 % of the total soluble protein in different transgenic lines (Guo et al. 2012b). Although the expression level of the protein is low, the results demonstrated that a forage plant-derived bioreactor system that represents a practical approach for production of avian influenza HA antigen protein.

For forage plant-based vaccines, a sufficient dose of antigen protein is generally required to ensure immunity efficiency. In all those cases, however, seemed to be relatively poor and could be further improved. Till date, recent advances have suggested several important avenues to optimize transgene expression at various steps that include: choice of promoter, gene integration and transcription, codon optimization, and protein accumulation (Guo et al. 2012b). In addition, a careful choice of plant host system could also be important for improving protein concentration.

10.5 Future Prospects

The ongoing developments in new sequencing technologies and methods will be crucial for future attempts to understand the genome structure and function of forage and turf grasses, to mine the important gene resources, and to develop and use new generation molecular markers for molecular breeding. The establishment of efficient tools for gene function analysis is necessary based on both transformation-based and alternative approaches such as TILLING and construction of mutant library, which have been successfully used in the model plants. With the development of current gene discovery methods, more and more functional genes and gene regulatory elements are cloned and identified. Once the highly efficient genetic transformation system of forage species are established, novel and improved materials with desirable commercial characteristics will be generated by genetic engineering. This may be the future trend of forage biotechnology no matter we accept genetically modified forage or not like GM crops. What we should do is strictly risk assessment of transgenic forage before they are produced commercially. The international and multidisciplinary collaborations are essential for future development to attract public and private funding, and to share resources that are hard to handle for single research institutes or groups such as forage genome sequencing, constructing large-scale association mapping populations and TILLING populations.

Acknowledgements This work was supported by National Basic Research Program of China ("973", 2014CB138704), the National High Technology Research and Development Program of China ("863", 2011AA100209).

References

Bao AK, Wang YW, Xi JJ, Liu C, Zhang JL, Wang SM (2014) Co-expression of xerophyte *Zygophyllum xanthoxylum* ZxNHX and ZxVP1–1 enhances salt and drought tolerance in transgenic *Lotus corniculatus* L. by increasing cations accumulation. Funct Plant Biol 41:203–214

Chen SY, Huang X, Yan XQ, Liang Y, Wang YZ, Li XF, Peng XJ, Ma XY, Zhang LX, Cai YY, Ma T, Cheng LQ, Qi DM, Zheng HJ, Yang XH, Li XX, Liu GS (2013) Transcriptome analysis in sheepgrass (*Leymus chinensis*): a dominant perennial grass of the Eurasian Steppe. PLoS One 8(7):e67974

Chen SY, Cai YY, Zhang LX, Yan XQ, Cheng LQ, Qi DM, Zhou QY, Li XX, Liu GS (2014) Transcriptome analysis reveals common and distinct mechanisms for sheepgrass (*Leymus chinensis*) responses to defoliation compared to mechanical wounding. PLoS One 9(2):e89495

Cheng LQ, Li XX, Huang X, Ma T, Liang Y, Ma XY, Peng XJ, Jia JT, Chen SY, Chen Y, Deng B, Liu GS (2013) Overexpression of sheepgrass R1-MYB transcription factor *LcMYB1* confers salt tolerance in transgenic Arabidopsis. Plant Physiol Bioch 70:252–260

Dong J, Wang XM, Wang Z, Gao H, Sun GZ (2012) Cloning and analysis of dihydroflavonol reductase (DFR) gene from *Medicago sativa*. Acta Pratac Sin 21(2):123–132

Guo Q, Wang P, Ma Q, Zhang JL, Bao AK, Wang SM (2012a) Selective transport capacity for K^+ over Na^+ is linked to the expression levels of *PtSOS1* in halophyte *Puccinellia tenuiflora*. Funct Plant Biol 39:1047–1057

Guo QQ, Zhang ZL, Jiang SJ, Ma JT, Xue WT, Wu YM (2012b) Expression of an Avian Influenza Virus (H5N1) Hemagglutinin gene in transgenic *Lotus corniculatus*. Plant Mol Biol Rep 30:1117–1124

Hu K, Yan XF, Li D, Tang XM, Yang H, Wang Y et al (2013) Genetic improvement of perennial ryegrass with low lignin content by silencing genes of CCR and CAD. Acta Pratac Sin 22(5):72–83

Huang LK, Zhang XQ, Xie WG, Zhang J, Cheng L, Yan HD (2012) Molecular diversity and population structure of the forage grass *Hemarthria compressa* (Poaceae) in south China based on SRAP markers. Genet Mol Res 11(3):2441–2450

Huo S, Chen H, Zhu QH, Xie XM (2012) Cloning of a 4CL gene fragment and construction of a RNAi expression vector of *Pennisetum purpureum*. Acta Pratac Sin 21(1):296–301

Jia HL, Wang XM, Gao HW, Dong J, Wang YQ, Liu JN et al (2012) Cloning the gene of γ-tocopherol methyltransferase from alfalfa and expression analysis in adverse situations. Acta Pratac Sin 21(6):198–206

Jiang J, Yang BL, Xia T, Yu SM, Wu YN (2011) Analysis of genetic diversity of salt tolerant alfalfa germplasms. Acta Pratac Sin 20(5):119–125

Jiang LF, Zhang XQ, Ma X, Huang LK, Xie WG, Ma YM, Zhao YF (2013) Identification of orchardgrass (Dactylis glomerata L.) cultivars by using simple sequence repeat markers. Genet Mol Res 12(4):5111–5123

Li D, Su Z, Dong J, Wang T (2009) An expression database for roots of the model legume *Medicago truncatula* under salt stress. BMC Genomics 10:517

Li D, Zhang Y, Hu X, Shen X, Ma L, Su Z, Wang T, Dong J (2011) Transcriptional profiling of *Medicago truncatula* under salt stress identified a novel CBF transcription factor *MtCBF4* that plays an important role in abiotic stress responses. BMC Plant Biol 11:109

Li XX, Gao Q, Liang Y, Ma T, Cheng LQ, Qi DM, Liu H, Xu X, Chen SY, Liu GS (2013a) A novel salt-induced gene from sheepgrass, *LcSAIN2*, enhances salt tolerance in transgenic Arabidopsis. Plant Physiol Bioch 64:52–59

Li XX, Hou SL, Gao Q, Zhao PC, Chen SY, Qi DM, Lee BH, Cheng LQ, Liu GS (2013b) *LcSAIN1*, a novel salt-induced gene from sheepgrass, confers salt stress tolerance in transgenic Arabidopsis and rice. Plant Cell Physiol 54(7):1172–1185

Liu RX, Yu LQ, Zhang Y, Wu D, Yun JF (2010) Clustering and evaluation of alfalfa germplasms with different fall dormancy levels by RAPD markers. Acta Agrestia sinica 18(1):108–114

Liu J, Zhao Q, Yang ZM (2012) Identification of nine *Cynodon dactylon* varieties by SRAP molecular markers. Chin J grassland 4:21–25

Liu Z, Ma L, Nan Z, Wang Y (2013a) Comparative transcriptional profiling provides insights into the evolution and development of the zygomorphic flower of *Vicia sativa* (Papilionoideae). PLoS One 8(2):e57338

Liu Z, Chen T, Ma L, Zhao Z, Zhao PX, Nan Z, Wang Y (2013b) Global transcriptome sequencing using the Illumina platform and the development of EST-SSR markers in autotetraploid alfalfa. PLoS One 8(12):e83549

Ma X, Gu XY, Chen TT, Chen SY, Huang LK, Zhang XQ (2013a) Genetic relationships between Lolium (Poaceae) species revealed by RAPD markers. Genet Mol Res 12(3):3246–3255

Ma Q, Li XY, Yuan HJ, Hu J, Wei L, Bao AK, Zhang JL, Wang SM (2013b) *ZxSOS1* is essential for long-distance transport and spatial distribution of Na^+ and K^+ in the xerophyte *Zygophyllum xanthoxylum*. Plant Soil 374:661–676

Pang Y, Peel GJ, Sharma SB, Tang Y, Dixon RA (2008) A transcript profiling approach reveals an epicatechin-specific glucosyltransferase expressed in the seed coat of *Medicago truncatula*. PNAS 105(37):14210–14215

Pang Y, Wenger JP, Saathoff K, Peel GJ, Wen J, Huhman D, Allen SN, Tang Y, Cheng X, Tadege M, Ratet P, Mysore KS, Sumner LW, Marks MD, Dixon RA (2009) A WD40 repeat protein from *Medicago truncatula* is necessary for tissue-specific anthocyanin and proanthocyanidin biosynthesis but not for trichome development. Plant Physiol 151(3):1114–1129

Pang Y, Cheng X, Huhman D, Ma J, Peel JG, Yonekura-Sakakibara K et al (2013) Medicago glucosyltransferase UGT72L1: potential roles in proanthocyanidin biosynthesis. Planta 238(1):139–154

Peng XJ, Ma XY, Fan WH, Su M, Cheng LQ, Alam I, Lee BH, Qi DM, Shen SH, Liu GS (2011) Improved drought and salt tolerance of *Arabidopsis thaliana* by transgenic expression of a novel DREB gene from *Leymus chinensis*. Plant Cell Rep 30:1493–1502

Peng XJ, Zhang LX, Zhang LX, Liu ZJ, Cheng LQ, Yang Y, Shen SH, Chen SY, Liu GS (2013) The transcriptional factor *LcDREB2* cooperates with *LcSAMDC2* to contribute to salt tolerance in *Leymus chinensis*. Plant Cell Tiss Org 113:245–256

Rao JP, Agrawal P, Mohammad R, Rao SK, Reddy GR, Dechamma HJ, S Suryanarayana VV (2012) Expression of VP1 protein of serotype A and O of foot-and-mouth disease virus in transgenic sunnhemp plants and its immunogenicity for guinea pigs. Acta Virol 56(2):91–99

Shen X, Martens, Chen M, Li D, Dong J, Wang T (2010) Cloning and characterization of a functional flavanone-3-hydroxylase gene from *Medicago truncatula*. Mol Biol Rep 37(7):3283–3289

Su M, Li XX, Li XF, Cheng LQ, Qi DM, Chen SY, Liu GS (2013) Molecular characterization and expression analysis of a sheepgrass sucrose transporter *LcSUT1* after defoliation. Plant Mol Biol Rep 31(5):1184–1191

Sun Y, Long R, Kang J, Zhang T, Zhang Z, Zhou H, Yang Q (2013) Molecular cloning and characterization of three isoprenyl diphosphate synthase genes from alfalfa. Mol Biol Rep 40(2):2035–2044

Tang R, Zhang XQ, Li YH, Xie XM (2014) Cloning and in silico analysis of a cinnamyl alcohol dehydrogenase gene in *Pennisetum purpureum*. J Genet 93(1):145–158

Tong ZY, Xie C, Ma L, Liu LP, Jin YS, Dong JL, Wang T (2014) Co-expression of bacterial aspartate kinase and adenylylsulfate reductase genes substantially increases sulfur amino acid levels in transgenic alfalfa (*Medicago sativa* L.). PLoS One 9(2):e88310

Wang DM, Zhu JB, Peng M, Zhou P (2008) Induction of a protective antibody response to FMDV in mice following oral immunization with transgenic *Stylosanthes* spp. as a feedstuff additive. Transgenic Res 17(6):1163–1170

Wu GQ, Xi JJ, Wang Q, Bao AK, Ma Q, Zhang JL, Wang SM (2011) The *ZxNHX* gene encoding tonoplast Na^+/H^+ antiporter from the xerophyte *Zygophyllum xanthoxylum* plays important roles in response to salt and drought. J Plant Physiol 168:758–767

Xie WG, Zhang XQ, Cai HW, Liu W, Peng Y (2010) Genetic diversity analysis and transferability of cereal EST-SSR markers to orchardgrass (*Dactylis glomerata* L.). Biochem Syst Ecol 38(4):740–749

Xie WG, Zhang XQ, Cai HW, Huang LK, Peng Y, Ma X (2011) Genetic maps of SSR and SRAP markers in diploid orchardgrass (*Dactylis glomerata* L.) using the pseudo-testcross strategy. Genome 54(3):212–221

Xie C, Zhang R, Qu Y, Miao Z, Zhang Y, Shen X, Wang T, Dong J (2012) Overexpression of *MtCAS31* enhances drought tolerance in transgenic Arabidopsis by reducing stomatal density. New Phytol 195(1):124–135

Zhang XY, He JG, Sun XH, Wu JX (2011) Transformation of *Lolium perenne* with a fructan: fructan 1-fructosyltransferase gene from *Agropyron cristatum* and enhancement of drought tolerance in transgenic plants. Acta Pratac Sin 20(1):111–118

Chapter 11
Yield Components of Three *Brachiaria* Species Harvested at Two Cutting Frequencies in Hueytamalco, Puebla, Mexico

B. Alarcon-Zuniga, J. J. M. Ramirez-Gonzalez, S. Quijano-Hernandez and T. Cervantes-Martinez

11.1 Introduction

One of the strengths of livestock production systems in the tropics and subtropics is the forage genetic resources and grazing production capacity, where most livestock is raised up (Morgan 2010). In the subtropical region of the Sierra Madre Oriental of Mexico, particularly located in the states of Puebla, Veracruz and Oaxaca, native induced grasses (*Paspalum* and *Axonopus*) have persisted in considerable areas devoted to livestock, due to its ability to tolerate hills with low fertility soils, even though its forage yield, nutritional quality and persistence are very marginal for livestock production (Méndez 1999). The Sierra Madre Oriental is also characterized by a seasonal pasture growth: 1) a rainy season from June to October, when rainfall and warm climates increase forage growth; 2) a winter season, from November to February also called "nortes", which reduce growth of most forage species due to cloudy skies and low temperatures; 3) a drought season, from March to May, when the plant productivity decreases by erratic rainfall (Martinez et al. 2008). This climatological profile determines the persistence and consequent optimal stocking rate of native grassland in the Sierra Madre Oriental pastures. Other identified alternatives to improve livestock production is by selecting improved grasses, highly adapted to similar soil and climate conditions, but never probed in the Sierra Madre Oriental. *Brachiaria* genus has widely used in the lowland and highland American tropics, and includes about 100 species

B. Alarcon-Zuniga (✉) · S. Quijano-Hernandez · T. Cervantes-Martinez
Department of Animal Science, Universidad Autonoma Chapingo, Km 38.5. Hwg Mexico-Texcoco, Chapingo, Mex 56230, Mexico
e-mail: balarcon@correo.chapingo.mx

J. J. M. Ramirez-Gonzalez
Instituto Nacional de Investigaciones Forestales Agrícolas y Pecuarias, Experimental Field Station—Golfo Centro, Hueytamalco, Pue 73588, Mexico

H. Budak, G. Spangenberg (eds.), *Molecular Breeding of Forage and Turf,*
DOI 10.1007/978-3-319-08714-6_11

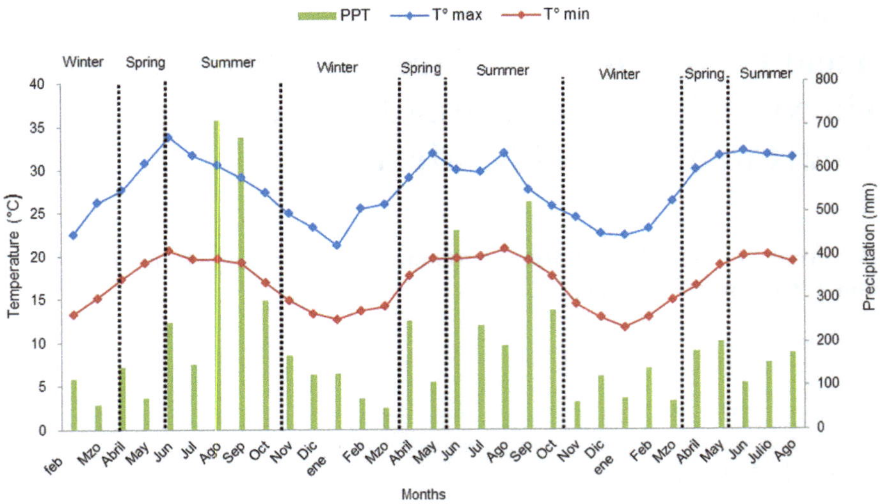

Fig. 11.1 Monthly temperature and rainfall recorded from February 2011 to August 2013 in the experimental field station "Las Margaritas" INIFAP Hueytamalco, Puebla, Mexico

highly adapted to hilly subtropical regions of Central and South America, and have a great ability to adapt to low fertility soils of hilly with a high content of aluminum, phosphorus and low calcium (Martinez et al. 2008). The most used species in this genus are *B. brizantha*, *B. decumbens*, *B. humidicola*, *B. ruziziensis* and *B. dyctioneura* (Faria 2006; Ruiz 2009). However, information is still lacking about adaptation, persistence and forage yield of *Brachiaria* species for each agro-climatic region of Mexico and particularly to the Sierra Madre Oriental. The objective was to determine the forage productivity of three *Brachiaria* species in the hilly mountain region of Puebla evaluated at different cutting frequencies in each season and their environmental interactions, assuming that both frequency and harvest intensity are aspects to consider in pasture management to maximize forage production and persistence (Sosa et al. 2008).

11.2 Material and Methods

As part of the collaborative work of the Mexican Breeding Initiative of Forage Crops, Chapingo Autonomous University and the National Institute of Forestry, Agricultural and Livestock Research of Mexico (INIFAP), the study was conducted in the experimental field "Las Margaritas" INIFAP, in Hueytamalco, Puebla, located at 19° 55' 25.75" N and 97° 17' 04.76" W, 896 msls, temperature of 21.5 °C and rainfall of 2714.1 mm. The distribution of rainfall is 69.5 % from June to mid-October, 18.1 % from October to mid-February called "North" or winter season; and 12.4 % from February to mid-June, drought season (Fig. 11.1). The experiment was located in the Sierra Madre Oriental, with hilly lands with 10–15 % slopes. Soils are characterized as Ultisols and Oxisols with clay loam texture—sandy, light brown to dark brown and soil pH is 4.5–5.5, considered as acid.

The work lasted two years, beginning on May 23, 2011 and ending on April 22, 2013. The experimental design was a randomized complete block split-plot arrangement and repeated measurements over time, which were distributed randomly in four replications; with major plots as *Brachiaria* species, and minor plots representing the harvesting frequencies, subdivided by season and years. ANOVA was computed using the GLM and MIXED procedures and Tukey mean comparison using SAS (2001). Forage crops evaluated were *Brachiaria humidicola* cv Chetumal (CH), *Brachiaria brizantha* cv Insurgente (IN) and *Brachiaria brizantha x Brachiaria ruziziensis* cv Mulato (MU). The harvest frequencies were done at 35 and 42 days, with total of 20 and 16 harvests, respectively. The experiment was uniformly cut at 10 cm in cv Chetumal and 20 cm in cvs Mulato and Insurgente, on April 17, 2011. It should be noted that to avoid adverse effects between adjacent plots, the harvest frequencies were arranged in ascending order starting with 35 days and ending on 42 days at the opposite end. If sufficient biomass to be cut at any of the frequencies mentioned were not accumulated during regrowth interval between harvests, then it was considered as "not sampling". The plots were 30 m^2 (3 m wide x 10 m long, including edges) with fertilization doses of 100-50-00, and nitrogen divided on two applications.

The experimental period covered three climatic seasons well distributed in the Sierra Madre Oriental of Mexico, which were considered as the rainfall season (RS) from June 15 to October 14; winter season (WI, also called Nortes) from October 15 to February 14 and the drought season (DR) from February to June 14, with two years of evaluation (2011 and 2012).

Average plant height, plant cover, total yield and leaf to stem ratio were measured. Plant height was measured for eight plants from floor to top without stretching inflorescence or leaves. At harvest time, Chetumal was cutoff at 10 cm and Mulato and Insurgentes at 20 cm. In the species being a tufted, cutting height was 10 cm and the semi-bunch species as cv Mulato or cv Insurgente and the height was 20 cm. Aerial coverage was measured to the visual assessment of the percentage of ground covered by the canopy of 1 m^2 sampling area. Total harvested biomass per plot was weighed and two sub-samples were taken to determine dry matter with 55 °C at 48 h, and fresh leaf to stem ratio.

11.3 Results and Discussions

The analysis of variance with the GLM and MIXED procedure showed that there were significant differences for forage yield among *Brachiaria* species, harvest frequencies, years and seasons ($P \leq 0.05$), and shown up in Table 11.1. These data are different as presented by Homen et al. (2010), where the highest forage yield were for *B. decumbens* and *B. humidicola* under conditions for tropical rainforest, but higher forage yield of the present study. Total coverage was higher for cv. Insurgent

Table 11.1 Dry matter yield (kg DM ha-1) leaf to stem ratio (% leaf: stem%), height (cm) and coverage (%) by cutting average of three species of Brachiaria spp. in Hueytamalco, Puebla

Cultivar	Forage yield components			
	Total forage yield, kg DM/ha	Leaf to stem ratio	Plant height	Cover
Chetumal	1882.4a[a]	77:23	38.7b	66c
Mulato	1895.8a	86:14	45.7a	67b
Insurgente	1940.6a	80:20	47.0a	69a

[a] Means with same letter within columns are not statistically different (p>0.05)

(IN), followed by cv Mulato (MU) and cv Chetumal (CH) with 69, 67 and 66%, respectively. As expected, the leaves percent was higher in cv Chetumal and Mulato with 86 and 77%, respectively, and considered as excellent leaf percent. These results differ from those reported by Gomez et al. (2000) in 24 accessions of *Brachiaria* where higher values were for *B. brizantha* and lower for *B. humidicola* in the Colombian Amazon Piedmont

The harvest frequency was highly significant ($P \leq 0.05$) on dry matter yield, plant height and coverage (Fig. 11.2), being higher at 42 days of regrowth (2383.8 kg DM/ha, 47 cm and 74%, respectively) in comparison to 35 days (1532.5, 41 and 63), similar as reported by Peralta et al. (2007); however, the cultivars were lower in leaf percent at 42 than at 35 days, as a result of stems growth (76 and 24%). This is because the L:S ratio of grasses is inversely proportional to age (Mancilla 2006, Peralta et al. 2007; Homen et al. 2010; Vergara and Araujo 2006). In regard to plant height, the results indicate that as the frequency is extended, plant height increases, this is because at 28 days after harvest, stems accumulate more in *Brachiaria* and leaves start decreasing and senescence, as mentioned by Homen et al. (2010) for *Brachiaria* species in Miranda, Venezuela.

The results of forage yield by seasons indicated that in rainy seasons, it accumulated more forage, followed by dry and winter, with 2871, 1619 and 1143 kg DM ha^{-1}, respectively ($P \leq 0.05$), indicating that environmental conditions of the hilly highland of Sierra Madre Oriental of Mexico during winter includes low temperatures, low and frequent rainfall, and cloudy skies, affects more than even high temperatures and low precipitation during the dry season. In the tropical lowland of Mexico (0–400 masl), during winter forage yield is still higher than dry season, and in some tropical grasslands of Tabasco, Campeche and southern Veracruz, forage yield is higher during dry season due to less cloudy skies than both rainfall and winter seasons (Mata 1989; Ramírez et al. 2009) Table 11.2.

The leaves percentage was lower during the rainy season and similar in winter and dry seasons, with 75 vs. 85 and 83%, respectively. Homen et al. (2010) and Cruz et al. (2011) reported an increase in leaves during the rainy season at lowland grasslands, which indicates that not only rainfall affects forage yield and leaf to stem ratio, but also low temperatures and cloudy skies frequently observed in the highlands of Sierra Madre Oriental of Mexico.

No differences were found within harvest frequencies per cultivar ($P \leq 0.05$), and at 35 days cvs Mulato, Chetumal and Insurgentes had 1656.2, 1507.5 and

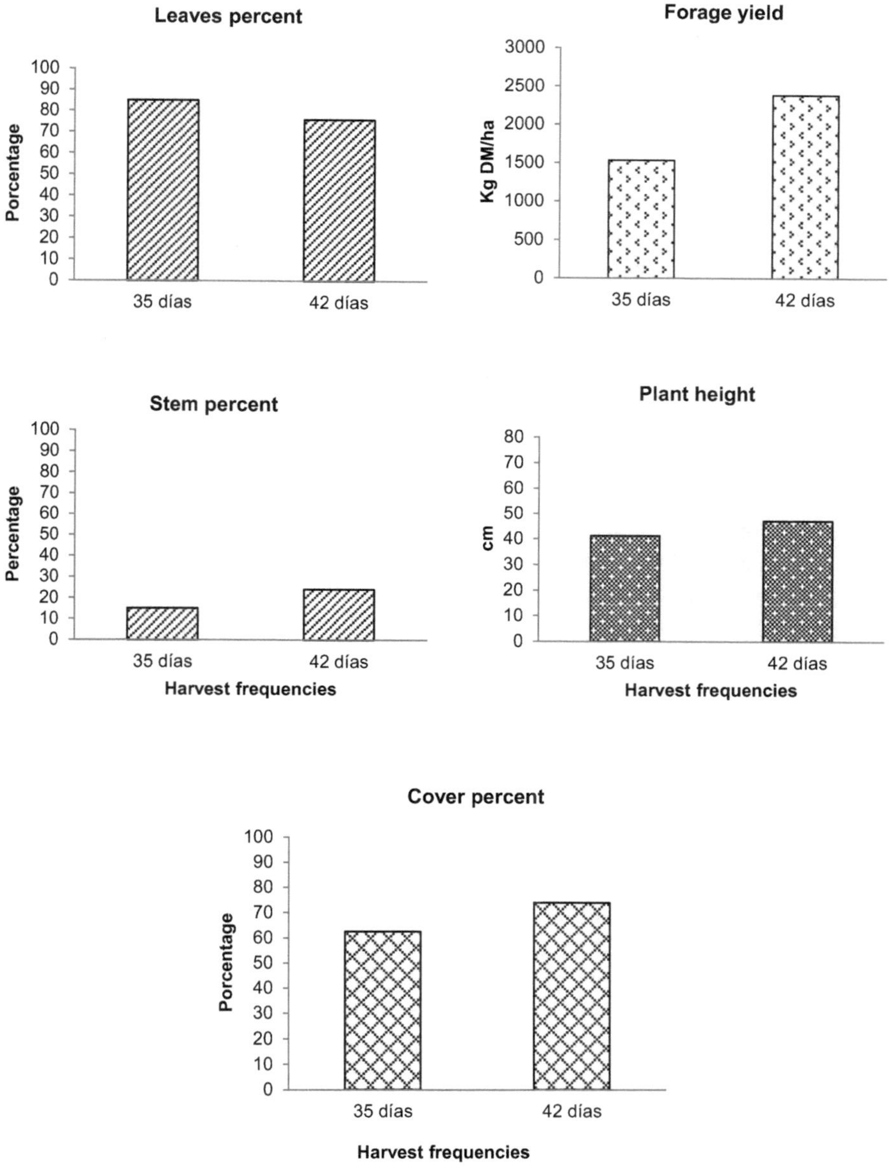

Fig. 11.2 Forage yield components of three species *Brachiaria spp*. Evaluated at two harvest frequencies in Hueytamalco, Puebla, Mexico

1434.0 kg DM ha^{-1}, respectively. Forage yield increase per unit area from 35 to 42 days was 58 % for Chetumal, 33 % in Mulato and 79 % for Insurgentes, indicating a positive trend to increase dry matter yield from 35 to 42 days, because the older leaf area is still increasing enough to accumulate forage biomass. This is consistent with reports of Enriquez (1997), who mentioned that dry matter production increases as

Table 11.2 Seasonal accumulation, morphological components and forage coverage of three *Brachiaria* species in Hueytamalco, Puebla, Mexico

	Seasons		
Item	Rainy	Winter	Drought
Forage yield, Kg DM ha^{-1})	2871a*	1143c	1619b
Leaves, %	75b	85a	83a
Stems, %	25a	15b	17b
Plant height, cm	57a	32c	42b
Coverage, %	77a	61c	65b

* Means with same letter within columns are not statistically different ($P>0.05$)

age advances. The plant height among species was higher at 42 days, with Insurgentes having the highest plant height for both frequencies (44 and 51 cm), followed by Mulato (43 and 49 cm) and Chetumal (36, 42 cm), similar as reported by Gomez et al. (2000). In this regard, Ramirez et al. (2009) mentioned that a greater harvest frequency, plant height and forage yield increase due to higher stem accumulation and dead material deposited on the ground, which reduces the leaf to stem ratio and alters morphological structure of the total forage yield, as occurs in cv Insurgent (Avellaneda et al. 2008).

Finally, coverage did not differ among cultivars ($P\leq 0.05$), but was lower at 35 days vs. 42 days, with an increase of 12% in last harvest frequency, and similar as reported by Canchilla et al. (2010). The interaction species by coverage did not show differences within seasons, but rainy season was more covered than winter and dry seasons, and this is similar to that found in Chetumal, Quintana Roo, and Guerrero, Mexico, with diverse grasses evaluated, forage yield and coverage performance was similar among forage crops at every season (Sosa et al. 2008; Rojas et al. 2011; Sanabria et al. 1995). The interaction of plant height and coverage per season were slightly lower than those achieved at 42 days for all time values. Homen et al. (2010) reported an average plant height higher than those found in this study. In the winter season, the stems percent was lower during winter at 35 days and correlated with coverage, which is a desirable feature in tropical forage crops. So then, during the dry season the leaf appearance rate and stems elongation are lower as the plant grows, and this performance is higher in rainy than dry seasons (Reynoso et al. 2010).

Low forage yields in the winter are attributed to climatic conditions from November to February (12 to 14 °C; cloudy skies), below those required for maximum yield potential of *Brachiaria* species, in comparison to the rainy season. In this regard, Moore (1981) mentions that the amount of forage yield produced is affected when there is occurrence of minimum temperatures below 22 °C (Alger 2005). At dry season, higher forage yield performance can be obtained at 42 days, compared to rainy season with best yield at 35 days. Machado (1996) mentions that the soil moisture deficit is affecting grasses growth, causing reduction in leaf area, photosynthetic activity and therefore the accumulation of forage.

11.4 Conclusions

Brachiaria species can be an excellent forage crop for the grassland of Sierra Madre Oriental of Mexico. Forage yield components were highly affected by season and harvest frequency, concluding than *Brachiaria brizantha* cv. Insurgente obtained the highest forage yield during rainy and dry seasons. At winter, *Brachiaria hunmidicola* cv. Chetumal can be excellent forage harvested at 35 days, and *Brachiaria* hybrid cv. Mulato was not well adapted to the highlands of Sierra Madre Oriental.

References

Argel PJ (2005) Contribución de los forrajes mejorados a la productividad Ganadera en sistemas de doble propósito. In: XIX Reunión de la Asociación Latinoamericana de Producción Animal. Tampico, Tamaulipas, México. pp. 42–50

Avellaneda CJ, Cabezas GF, Quintana ZG, Luna MR, Montañez VO, Espinoza GI, Zambrano MS, Romero GD, Vanegas RJ, Pinargote ME (2008) Comportamiento agronómico y composición química de tres variedades de *Brachiaria* en diferentes edades de cosecha. Quevedo, Ecuador. Ciencia y Tecnología 1(2):87–94

Canchilla ER, Soca M, Ojeda F, Machado R, Canchila N (2010) Dinámica de crecimiento de 24 accesiones de Brachiaria spp. Pastos y Forrajes [on line] 33(4):1–1. ISSN 0864–0394

Cruz LPI, Hernandez GA, Enriquez QJF, Mendoza PSI, Quero CAR, Joaquín TBM (2011) Desempeño agronómico de genotipos de *Brachiaria humidicola* (Rendle) Schweickt en el trópico húmedo de México. Rev Fitotec Mexicana 34(2):123–131 (Sociedad Mexicana de Fitogenética, A.C. Chapingo, México)

Enriquez QJF (1997). Evaluación agronómica de 13 cultivares de *Panicum maximum* en suelos ácidos del sur de Veracruz. XXXIII Reunión de Investigación Pecuaria en México. Veracruz, Ver. p. 13

Faría MJ (2006) Manejo de pastos forrajeros en la ganadería de doble propósito. X Seminario de Pastos y Forrajes. Posgrado de Producción Animal. Facultad de Agronomía. Universidad de Zulia, Maracaibo, Venezuela. p. 9

Gómez MM, Velásquez JE, Miles JW, Rayo FT (2000) Adaptación de *Brachiaria* en el Piedemonte Amazónico Colombiano. Nota de Investigación. Pasturas Tropicales 22(1):19–25

Homen M, Entrena I, Arriojas L (2010) Biomasa y valor nutritivo de tres gramíneas forrajeras en diferentes períodos del año en la zona de bosque húmedo tropical, Barlovento, estado Miranda. Zootecnia Trop 28(1):115–127. ISSN 0798–7269

Machado R (1996) Dinámica de algunos indicadores morfológicos y estructurales de *Andropogon gayanus* CIAT-621, bajo condiciones de manejo intensivo. Efecto de la época y el año. Pastos y Forrajes 19:121–127

Mancilla LE (2006) Manejo del pastoreo en la agricultura forrajera. X Seminario de Pastos y Forrajes. Universidad Nacional Experimental de los Llanos Occidentales. Ezequiel Zamora, UNELLEZ, Guanare. P.p. 24

Martínez MD, Hernández GA, Enríquez QJF, Pérez PJ, González MSS, Herrera HJG (2008) Producción de forraje y componentes del rendimiento del pasto *Brachiaria humidicola* CIAT 6133 con diferente manejo de la defoliación. Tec Pecu Mex 46(4):427–438

Mata PD (1989) Rendimiento y composición química de seis gramíneas introducidas en una sabana del sureste del estado Guarico. Zootec Trop 7(1):69–80

Méndez CJL (1999) Producción de materia seca de ocho gramíneas para pastoreo a diferentes edades y épocas de corte en Hueytamalco, Puebla. Tesis Profesional. Escuela de Ingeniería Agrohidráulica. Benemérita Universidad Autónoma de Puebla. San Juan Acateno, Teziutlán, Puebla. p. 62

Morgan E (2010) Tropical grasslands (Biomes Atlases). 2nd edn. Raintree Pub. 64 pages

Moore P (1981) The varied way plants tap the sun. New Scienetist 12:394–397

Peralta MA, Carrillo PS, Hernández HH, Porfirio PN (2007) Características morfológicas y productivas, en etapa de producción, para ocho gramíneas forrajeras tropicales. APPA—ALPA. Cusco, Perú. p. 3

Ramírez RO, Hernández GA, Carneiro SS, Pérez PJ, Enríquez QJF, Quero CAR, Herrera HJG, Cervantes NA (2009) Acumulación de forraje, crecimiento y características estructurales del pasto Mombaza (*Panicum máximum Jacq.*) cosechado a diferentes intervalos de corte. Tec Pecu Mex 47(2):203–213

Reynoso RO, Hernández GA, Carneiro da SS, Pérez PJ, Jacaúna de SJS, Castro RR, y Enríquez QJF (2010) Características morfogénicas y su influencia en el rendimiento del pasto Mombaza, cosechado a diferentes intervalos de corte. Trop Subtrop Agroecosyst 12(2):303–311

Rojas HS, Olivares PJ, Jiménez GR, Gutiérrez SI, Avilés NF (2011) Producción de materia seca y componentes morfológicos de cuatro cultivares de *Brachiaria* en el trópico. Avances en Investigación Agropecuaria

Ruiz DC (2009) Evaluación de componentes de rendimiento y calidad nutricional de Brachiaria brizantha a diferentes dosis y fuentes de nitrógeno en Villa Ávila Camacho, Puebla. Tesis Profesional. Departamento de Zootecnia. Universidad Autónoma Chapingo. p. 67

Sanabria VDM, Fariñas JG, Manrique U, Flores Z, Reina Y (1995) Adaptabilidad de Gramíneas y Leguminosas forrajeras en un paisaje de Mesa del estado Bolívar, Venezuela. Zootec Trop 13(1):63–76

Sosa RE, Cabrera TE, Pérez RD, Ortega RL (2008) Producción estacional de materia seca de gramíneas y leguminosas forrajeras con cortes en el estado de Quintana Roo. Tec Pecu Mex 46(4):413–426

Vergara LJ, Araujo FO (2006) Producción, composición química y degradabilidad ruminal in situ de Brachiaria humidicola (Rendle) Schweick en el bosque seco tropical. Rev Cient Venezuela 16(3):28–40

Chapter 12
Use of SSR Markers to Increase Forage Yield in Timoty (*Phleum pratense* L.)

Tsuneki Tanaka, Hiroyuki Tamaki, Kazunori Ashikaga, Hiroki Fujii, Ken-ichi Tamura and Toshihiko Yamada

12.1 Introduction

Timothy (*Phleum pratense* L.), a perennial and cross-pollinating hexaploid ($2n=6x=42$) species with a high degree of self-incompatibility, is an important perennial forage grass in Nordic countries, eastern Canada and northern Japan (Tamaki et al. 2010). Raising the forage yield remains a high-priority target of timothy breeding in Japan (Ueda 1990; Tamaki et al. 2010). However, forage yield increases of timothy have lagged behind those of many other crops (Tamaki et al. 2009). The same limited yield improvement has been reported for other forage grasses or legumes in which amount of improvement was small despite over 100 years of breeding efforts (Casler and Brummer 2008). One of the major reasons for this stagnation is that the primary breeding strategy in such forage grasses or legumes, with perfect flowers and without a stable cytoplasmic male sterility system, is still variety synthesis based on general combining ability (GCA) mainly due to additive effect (Tamaki et al. 2007). A potentially effective way to improve forage yield in

T. Yamada (✉)
Field Science Center for Northern Biosphere, Hokkaido University,
Kita 11 Nishi 10, 060-0811 Sapporo, Hokkaido, Japan
e-mail: yamada@fsc.hokudai.ac.jp

T. Tanaka · K. Ashikaga · H. Fujii
Hokkaido Research Organization, Kitami Agricultural Experiment Station,
52 Yayoi, Kunneppu, Tokoro-gun, 099-1496 Sapporo, Hokkaido, Japan

H. Tamaki
National Institute of Livestock and Grassland Science,
768 Senbonmatsu, Nasusiobara, 329-2793 Tochigi, Japan

K. Tamura
National Agricultural and Food Research Organization, Hokkaido Agricultural Research Center,
1 Hitsujigaoka, Toyohira, 062-8555 Sapporo, Hokkaido, Japan

H. Budak, G. Spangenberg (eds.), *Molecular Breeding of Forage and Turf*,
DOI 10.1007/978-3-319-08714-6_12

the future is to capture heterosis or exploit specific combining ability (SCA) mainly due to nonadditive effects (Li and Brummer 2009). Several breeding methods have been proposed for effective yield improvement in forage grasses or legumes with the aim of capitalizing on nonadditive effect (Brummer 1999; Tamaki et al. 2007, 2009). One of the methods is a "clone and strain synthesis (CSS)" approach that exploits both GCA and SCA and minimizes the risk of inbreeding depression (Tamaki et al. 2009). The scheme, which is similar to variety synthesis, consists of two procedures: (i) crossing two seed parental clones (SPs) with a common pollen parental strain (PP) regarding the seeds harvested from the two SPs as first synthetic (Syn1) generation, that is, developing two top-cross progenies, and (ii) even mixing of the Syn1 seeds (balanced bulk of two top-cross progenies) to produce subsequent generations, as is done to develop conventional synthetic varieties from a number of parental clones (Tamaki et al. 2009). This has been used to develop strains producing significantly higher yield performance of the second synthetic (Syn2) generation (Tamaki et al. 2009) and the forth synthetic (Syn4) generation corresponding to commercial seeds (Tanaka et al. 2013a) in timothy.

Molecular markers are regarded as powerful tools in forage grasses for the analysis of genetic diversity and the identification of parents for variety developments capitalizing heterosis or avoiding an inbreeding depression (Kölliker et al. 2005; Amini et al. 2011). Attempts of relationships between marker-based genetic distances (GDs) and forage yields were reported in forage grasses (Kölliker et al. 2005; Amini et al. 2011; Robins et al. 2012) or legumes (Kidwell et al. 1994, 1999; Joyce et al. 1999; Tucak et al. 2011). Although failed study has been reported in alfalfa (Kidwell et al. 1999), molecular markers are particularly helpful in identifying genetically similar germplasm when large germplasm collections exist but no heterotic groups are available (Reif et al. 2003). In timothy, significantly positive correlations between marker-based GD using 28 simple sequence repeat (SSR) markers and forage yields were found in polycross progenies (Tanaka et al. 2013b), top-cross progenies (Tanaka et al. 2011a) and CSS strains (Tanaka et al. 2011b). These results indicated that marker-based GD could be used as a tool for predicting heterosis or SCA and that considerable SCA effect contributed to forage yields in timothy.

The strategy exploiting heterosis in timothy CSS breeding programs requires an establishment of reciprocal recurrent selection (RRS) scheme using genetic diversity based on DNA markers. Information on relative contribution of GCA and SCA help breeders devise breeding and selection strategies. The aims of this study were (i) to examine the relative contribution of GCA and SCA to forage yield in CSS breeding scheme by predicting yields of Syn1 and Syn2 with molecular marker and field evaluation data, and (ii) to propose feasible RRS scheme with CSS development based on results obtained in this study.

12.2 Methods

12.2.1 Estimation of GCA of Parental Materials

Dataset of 100 entries examined in 11 polycross progeny tests at Kitami Agricultural Experiment Station (KAES) during 1971–2013 were used in this experiment. The 100 entries included 5 early maturing check varieties ("Climax," "Hokuo," "Senpoku," "Nosappu," and "Aurora"), which were examined over multiple tests. The other 95 entries included 50 polycross progenies derived from clones reported by Tanaka et al. (2011a), which are the representatives of the early maturing breeding populations at KAES. In each polycross progeny test, the means of annual dry matter yields (DMYs; Mg ha^{-1}) over 2 or 3 years except for seedling-year were examined. The broad-sense heritability (h_B^2) of the annual DMYs in each polycross progeny test was estimated based on the *F*-value of the entries in analysis of variance (ANOVA) as follow: $h_B^{2}=1-1/F$-value. The random effects of the entries were estimated by best linear unbiased prediction (BLUP) based on the annual DMYs according to the mixed effect model as following equation: $y=X\beta+Z\alpha+\varepsilon$, where y is the vector of annual DMYs, β and α are vectors of fixed and random effects, respectively, X and Z are the associated design matrices, and ε is a random residual vector. The fit of the model by an iterative process gives estimates of the variance components, a prediction of random vector and a solution for the fixed effect vector. The process most used to estimate the components of variance is restricted/residual maximum likelihood (REML), which was developed by Patterson and Thompson (1971). Heritability (h^2) among half-sib families was estimated according to the expression: $h^2=\sigma^2_G/(\sigma^2_G+\sigma^2_E)$, where σ^2_G is random effect variance and σ^2_E is residual variance, respectively, in the REML process.

12.2.2 Relative Contribution of General and Specific Combining Abilities in Forage Yields of Top-Cross Progenies

Timothy 51 top-cross progenies, which correspond to the Syn1 generation of CSS, and their SPs were used in this experiment. The cumulative DMYs over 2 years (2003–2004) of the progenies and marker-based GDs of SPs toward a PP (GD_T) were already examined in earlier study (Tanaka et al. 2011a). Field trails and genotyping protocol were described by Tanaka et al. (2011a). In addition, GCA of the 51 SPs was estimated by the BLUP in this study (Table 12.1), except for "a-13" which had not been examined in polycross progeny tests. Effective progeny numbers (n_E) of SPs (Table 12.1) indicating connectedness of data in BLUP procedure were calculated as follows (Robertson and Rendel 1954): $n_{Ei}=\Sigma\{n_{i,j}(1-n_{i,j}/n_j)\}$, where $n_{i,j}$ is a number of *i*th entries in *j*th progeny test ($n_{i,j}=1$ in all case of this experiment), n_j

Table 12.1 Design matrix[a] in 11 timothy (*Phleum pratense* L.) polycross progeny tests (A–K), the effective progeny number (n_E)[b], and general combining ability (GCA)[c] for annual dry matter yields (DMYs) of 50 parental clones

Parental clone designations	Entry of polycross progeny tests												
	A	B	C	D	E	F	G	H	I	J	K	n_E	GCA
b-04	Y								Y			1.87	0.162
b-08	Y						Y					1.87	0.157
b-12										Y		0.83	0.103
b-15	Y											0.97	0.378
b-31	Y											0.97	0.367
b-40						Y						0.90	0.039
c-09						Y						0.90	−0.027
c-33									Y			0.90	−0.075
d-09						Y						0.90	0.025
d-39							Y					0.90	−0.016
a-05									Y			0.90	−0.074
a-06									Y			0.90	−0.029
a-22							Y					0.90	−0.010
a-25										Y		0.83	−0.081
b-24	Y				Y							1.88	−0.132
b-29	Y											0.97	0.036
c-08	Y						Y					1.87	0.269
c-13	Y											0.97	−0.281
c-24	Y										Y	1.80	0.124
c-34							Y					0.90	−0.087
d-17				Y								0.86	−0.187
d-28											Y	0.83	−0.153
a-07								Y				0.88	0.081
a-10				Y								0.86	0.076
a-38											Y	0.83	0.514
a-40						Y						0.90	−0.069
b-09						Y						0.90	0.064
b-19								Y				0.88	−0.017
b-26	Y											0.97	0.250
b-32		Y	Y									1.95	0.409
b-37					Y							0.92	−0.063
c-04									Y			0.90	−0.013
c-17				Y								0.86	−0.109
c-23	Y				Y							1.88	0.330
c-36							Y					0.90	0.004
c-39	Y						Y					1.87	−0.049
d-03									Y			0.90	−0.039

Table 12.1 (continued)

Parental clone designations	Entry of polycross progeny tests												
	A	B	C	D	E	F	G	H	I	J	K	n_E	GCA
d-07								Y				0.88	0.055
d-10	Y				Y							1.88	0.028
d-23	Y				Y							1.88	0.336
d-24											Y	0.83	0.085
d-33								Y				0.88	0.011
a-01		Y										0.98	0.226
a-11					Y							0.92	−0.059
a-23	Y				Y							1.88	−0.181
a-37	Y				Y							1.88	0.168
b-11					Y							0.92	−0.030
b-22	Y					Y						0.97	−0.035
c-01	Y	Y										1.94	0.595
c-02	Y			Y								1.82	−0.175

[a] Y means entry to each polycross progeny tests: A–K

[b] The effective progeny number (n_E) is calculated according to Robertson and Rendel (1954)

[c] GCA is estimated by fitting to a mixed additive model based on annual DMYs in 11 polycross progeny tests with 180 data and 100 entries

is all number of targeted entries in *j*th progeny test. To investigate relative contributions of GCA and SCA to yields of Syn1 generation of CSS strains (i.e., top-cross progenies), the dataset including the cumulative DMYs over two years of top-cross progenies, GD_T and GCA, was analyzed according to the following multiple linear regression: $y=\beta_0+\beta_1\ GD_T+\beta_2\ GCA$, where y is the cumulative DMYs of top-cross progenies, β_0, β_1 and β_2 were partial regression coefficients (PRCs), GD_T was marker-based GDs toward a PP, and GCA was BLUP value based on polycross progeny tests, respectively. In addition, standardized PRCs and the Pearson product-moment correlation coefficients among the three data (DMY of top-cross progenies, GD_T, and GCA) were calculated.

12.2.3 Relative Contribution of Genetic Effects in Syn2 Generation of CSS Strains

Syn2 generation of 15 CSS strains reported by Tamaki et al. (2009) and their parental materials were used in this experiment. Seed parent clones of the 15 ($={}_6C_2$) CSS strains were "b-15," "c-01," "c-02," "c-08," "c-17," and "c-23," which were selected for high yields out of 152 entries in top-cross progeny tests described by Tamaki et al. (2009). The common PP of the CSS strains was "Aurora."

The cumulative annual DMYs of 15 CSS strains, which had been developed from all possible pairs of six top-cross progenies derived from crosses between the SPs and "Aurora," were evaluated at the field of the KAES, Hokkaido Research Organization, Kunneppu, Hokkaido, Japan, (43°47′N, 143°42′E) on a high-humic haplic wet andosol over 2 years (2005–2006). Field evaluations were described in earlier study by Tamaki et al. (2009). The six SPs and 16 individuals of "Aurora" as a PP were used in SSR evaluation. The outlines of the genotyping protocol used were described in Tanaka et al. (2011b). Genetic distance estimates were calculated from SSR data for all possible pairs of genotypes using the following equation: $GD_{i,j} = 1 - [2N_{i,j}/(N_i + N_j)]$, where $GD_{i,j}$ is the GD estimate between clone *i* and *j*, $N_{i,j}$ is the total number of bands common to clone *i* and *j*, and N_i is the total number of bands present in *i* (Dice 1945; Nei and Li 1979). Principle coordinate analyses were performed on the matrix of the GD estimates with the program GenAlEx (Peakall and Smouse 2006).

To investigate relative contribution of GCA and SCA to yields of Syn2 generation of CSS strains, data set including the cumulative DMYs of Syn2 of CSS strains, the cumulative DMYs of Syn1, sum of GCA (ΣGCA) of the combinations in the six SPs examined by BLUP, marker-based GD estimates among SPs, and marker-based GD between SPs and a PP (GD_T) was analyzed according to multiple linear regression analysis with selection of variables based on Akaike's information criterion (AIC) employing the function "lm" in the R statistical software (GNU General Public License).

12.2.4 Selection Simulations for RRS with CSS

Based on the grand mean and variance components in REML/BLUP procedure, the prediction models for yields of top-cross progenies and CSS strains, and 0.05 of the narrow sense heritability among individuals within families obtained by actual timothy breeding data in Japan, the potential of genetic gain of the forage yield for three patterns of RRS (Fig. 12.1) and among and within half-sib selection (AWHS) was estimated by selection simulation performed with R statistical software (GNU General Public License). The population size was set as 8000 individuals of 80 half-sib families (100 individuals per half-sib family), which was a conventional population size in timothy breeding at KAES. The selection pressures were 0.05 (among individuals within half-sib families) and 0.2 (among half-sib families), i.e., 80 superior individuals of 16 superior half-sib families were selected from 8000 individuals of 80 half-sib families. The genetic increases of forage yields through four selection cycles were calculated by generating arbitrary random numbers following the genetic parameters with 100 replications.

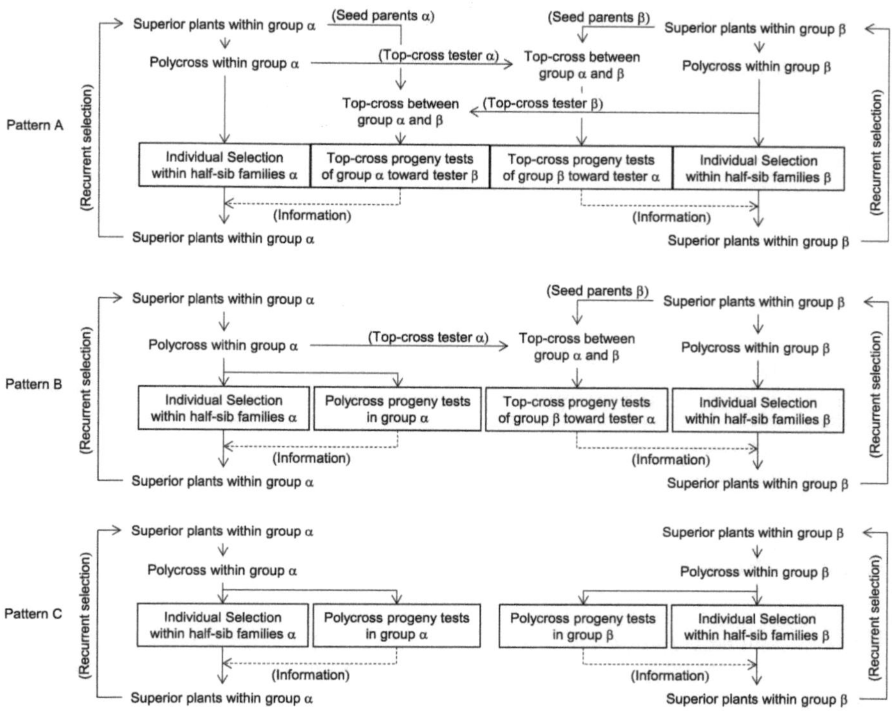

Fig. 12.1 Patterns (A–C) of individual selection within each heterotic group with evaluations of combining ability by test-cross and its polycross progeny test in reciprocal recurrent selection (RRS) procedures. Pattern A (*upper* part in the figure) shows procedures adopted for top-cross progeny tests in each group. Pattern B (*middle* part in the figure) shows procedures adopted for polycross and top-cross progeny tests in groups α and β, respectively. Pattern C (*lower* part in the figure) shows procedures adopted for polycross progeny tests in each group.

12.3 Results

12.3.1 Estimation of GCA of SPs

The grand mean was 11.1 Mg ha^{-1} in the dataset of the 11 polycross progeny tests. The BLUP values were −0.92–0.60 with mean of 0.0 Mg ha^{-1}. Variance components for forage yields of polycross progenies by REML were estimated as 0.115 (σ^2_G) and 0.150 (σ^2_E). These showed a moderate heritability estimate (h^2=0.43) among half-sib families. The GCA estimates of 51 SPs were shown in Table 12.1.

Table 12.2 Regression analysis of variance of marker-based genetic distances (GD_T) toward a tester and general combining ability (GCA) estimates of seed parents on cumulative annual dry matter yields (DMYs; Mg ha^{-1})[a] of 50 timothy (*Phleum pratense* L.) top-cross progenies over 2 years (2003–2004) at Kunneppu, Hokkaido, Japan

Source of variance	df	Mean of squares	*F*-value	*P*-value
Regression	2	29.8	32.6	<0.001
GD_T[b]	1	48.0	52.6	<0.001
GCA[c]	1	11.5	12.6	<0.001
Residual	47	0.914		
Total	49			

[a] Experimental plots were harvested three times per year during 2003–2004
[b] GD_T was calculated from the average of genetic distances (GDs) between each seed parent and 16 individuals of tester "Aurora" based on 28 simple sequence repeat (SSR) marker diversity
[c] Best linear unbiased prediction (BLUP) values of SPs in mixed effect models based on 11 polycross progeny tests with 180 data and 100 entries, where polycross progenies of the SPs were examined

12.3.2 Predicting Yields of Top-Cross Progenies

The correlations (r) among DMYs of 50 top-cross progenies, GD_T and GCA of their SPs were −0.02 ($p>0.05$, GD_T vs. GCA), 0.68 ($p<0.001$, DMY vs. GD_T), and 0.33 ($p<0.05$, DMY vs. GCA), respectively. The multiple linear regression analysis predicting DMYs of top-cross progenies with GD_T and GCA produced significant (F-value = 32.6; df = 2, 47; $p<0.001$) regression models (Table12.2). The respective PRCs (β_0, β_1 and β_2) in the regression analysis were −8.57 (β_0; intercept), 42.8 (β_1; the PRC of GD_T), and 2.69 (β_2; the PRC of GCA). The standardized PRCs of GD_T (0.68) was larger than that of GCA (0.35). The multiple correlation coefficient was 0.76. The squared multiple correlation coefficient adjusted for the degrees of freedom was 0.56.

12.3.3 Predicting Yields of CSS Strains

As results of the selection of the explanatory variables based on AIC in the multiple linear regression analysis, two factors which were marker-based GD among SPs and ΣGCA of the two SPs examined using BLUP were selected for prediction of the yield of Syn2 of CSS strains (Table 12.3). The correlations between response and explanatory variables were, respectively, 0.66 ($p<0.01$, DMY vs. ΣGCA) and 0.64 ($p<0.05$, DMY vs. GD). Analysis of variance of the regression (Table 12.3) found significance of regression (F-value = 13.8; df = 2, 12; $p<0.001$). The prediction model, in which all PRCs were significant, was the following equation: $y_{i,j} = 19.7 + 12.9\ GD_{i,j} + 2.40\ (GCA_i + GCA_j)$, where $y_{i,j}$ is DMY of Syn2 of CSS

Table 12.3 Regression analysis of variance of marker-based genetic distances (GDs) among timothy (*Phleum pratense* L.) seed parents (SPs) of "clone and strain synthesis (CSS)" strains and sum of general combining ability (GCA) estimates of SPs on cumulative annual dry matter yields (DMYs; Mg ha^{-1})[a] of 15 CSS strains over 2 years (2005–2006) at Kunneppu, Hokkaido, Japan

Source of variance	df	Mean of squares	*F*-value	*P*-value
Regression	2	11.9	13.8	0.001
GD[b]	1	9.12	10.6	0.01
GCA[c]	1	14.7	17.0	0.01
Residual	12	0.863		
Total	14			

[a] Experimental plots were harvested three times per year during 2005–2006
[b] GD were calculated based on 28 simple sequence repeat (SSR) markers diversity according to Dice (1945) and Nei and Li (1979)
[c] Best linear unbiased prediction (BLUP) values of SPs in mixed effect models based on 11 polycross progeny tests with 180 data and 100 entries, where polycross progenies of the SPs were examined

strain derived from SP *i* and *j*, $GD_{i,j}$ represents marker-based GD estimates between SP *i* and *j*, and GCA_i stands for the GCA estimate of SP *i* (Table 5.6). The standardized PRCs of GD (0.53) were similar to that of ΣGCA (0.55). The multiple correlation coefficient was 0.83. The squared multiple correlation coefficient adjusted for the degrees of freedom was 0.65.

12.3.4 Expected Genetic Gains for Forage Yields by RRS with CSS

Genetic gains for cumulative DMYs over 2 years through three selection cycles in the three schemes were 5.1 Mg ha^{-1} (22.3 %) in pattern A, 4.2 Mg ha^{-1} (19.3 %) in pattern B, and 2.2 Mg ha^{-1} (10.6 %) in pattern C (Fig. 12.2). Patterns A and B acquired twice the genetic gains as AWHS did, attaining one selection cycle. Pattern C was not so much compared with AWHS (Fig. 12.2). The increase of the additive effect through three selection cycles in the population of the pattern A established top-cross progeny tests (0.747 Mg ha^{-1}) was less than that in AWHS (1.19 Mg ha^{-1} through four selection cycles), whereas the increase in the pattern C established polycross progeny tests (0.915 Mg ha^{-1}) was nearly equal to that in AWHS. For pattern B, one group in the procedure of top-cross progeny tests was similar to that of pattern A. Another group in polycross progeny tests was equaled to that of pattern C.

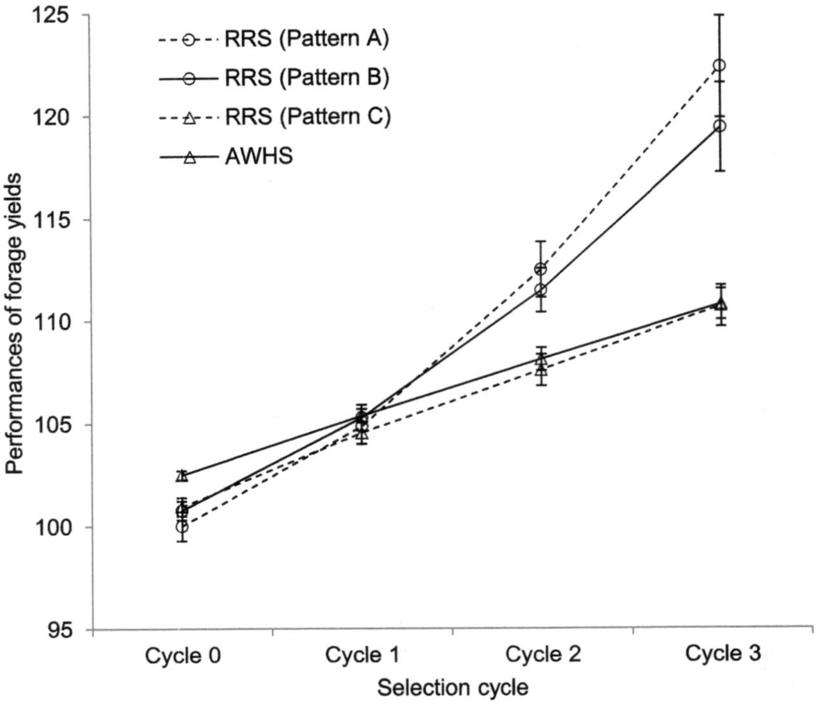

Fig. 12.2 Transition of forage yields with selection cycles of three patterns (A, B, and C) of reciprocal recurrent selection (RRS) with clone and strain synthesis (CSS) development and among and within half-sib selection (AWHS) in timothy (*Phleum pratense* L.) by computer simulation based on the dataset obtained in this study. Performances of forage yields (PY) in RRS were calculated as the yield of cycle 0 in the pattern A was equal to 100. PY in AWHS was calculated the yield of cycle 0 was equal to 100 on the assumption that selection cycle in AWHS had advanced one selection cycle than that in RRS, i.e., cycle 1 in AWHS corresponds to cycle 0 in this figure.

12.4 Discussion

Estimating the effects of GCA and SCA helps breeders to allocate resources for parental selections in CSS breeding scheme because various genetic effects such as GCA of parental materials (SPs and PP) and SCA among parental materials are expected to influence the forage yields of Syn1 and Syn2 of CSS strains, respectively. This study was conducted to examine the relative contributions of GCA and SCA to forage yield in CSS breeding scheme by predicting yields of Syn1 and Syn2 with molecular marker and field evaluation data. Multiple linear regression analysis (Table 12.2), used to predict yields of top-cross progenies, where standardized PRCs of GD_T between SPs and a tester as PP and GCA of SPs were 0.68 and 0.35, shows a greater contribution of SCA in yields rather than GCA. These results demonstrate that molecular marker diversity can be anticipated for wider use in the identification of different heterotic groups in a current timothy breeding population. Marker-based GDs among SPs and sum of GCA of two SPs examined using BLUP values (Table 12.1) were selected as the explanatory variables in the yield prediction models in Syn2 (Table 12.3), suggesting that utilization of putative heterotic groups as the first step in a RRS approach should be selection based on GCA of SPs and SCA among SPs. An adoption of elite clones as SPs of CSS strains, selected from putative heterotic groups formed by molecular markers, results in the efficient development of high-yielding strains with exploiting SCA among SPs. The CSS development presents some important advantages over conventional cultivar synthesis: RRS with two or three heterotic groups can be established in CSS; also, current leading cultivars are useful as the PP.

Posselt (2010) described a generalized RSS scheme for improving two parent populations in hybrid breeding. Feasible RRS approaches were selections of GCA of SPs and GD among SPs in this study. Taking these two perspectives into consideration, the breeding scheme was proposed, which consists of three breeding phases. In brief, current breeding populations were divided into two groups based on molecular marker diversity in the first phase of creation of the base population. In the second phase of parental selections, SP candidates were selected based on individual selections within each of the two pre-divided groups and top-cross progeny tests with a tester derived from another group. In the third phase of constructions and evaluation of experimental varieties, CSS strains were developed by combinations of SP candidates and leading varieties or well-developed strains as a PP.

On the assumptions that (i) only two (putative) heterotic groups are identified by genetic diversity on SSR makers and (ii) an initial PP was not improved because no evidence for the third heterotic group and feasible development of PP candidates was shown in this study, three patterns of RSS schemes combined procedures of either top-cross or polycross progeny tests were compared (patterns A–C in Fig. 12.1). Genetic gains for cumulative DMYs over 2 years through three selection cycle in the three schemes indicated that the pattern B of RRS is a more feasible scheme for a rapid yield increase avoiding the loss of accumulating additive effects within the populations. Operations of RRS with CSS development using genetic diversity on SSR markers are just launched in timothy breeding at KAES. In the future, these approaches are expected to achieve long-term rapid yield improvement in timothy.

Acknowledgments Funding was partially provided by a grant from the Northern Advancement Center for Science & Technology, Hokkaido, Japan.

Reference

Amini F, Mirlohi A, Majidi MM, ShojaieFar S, Kölliker R (2011) Improved polycross breeding of tall fescue through marker-based parental selection. Plant Breed 130:701–707

Brummer EC (1999) Capturing heterosis in forage crop cultivar development. Crop Sci 39:943–954

Casler MD, Brummer EC (2008) Theoretical expected genetic gains for among-and-within-family selection methods in perennial forage crops. Crop Sci 48:890–902

Dice LR (1945) Measures of the amount of ecological association between species. Ecology 26:297–302

Joyce TA, Abberton MT, Michaelson-Yeates TPT, Forster JW (1999) Relationships between genetic distance measured by RAPD-PCR and heterosis in inbred lines of white clover (*Trifolium repens* L.). Euphytica 107:159–165

Kidwell KK, Woodfield DR, Bingham ET, Osborn TC (1994) Molecular marker diversity and yield of isogenic 2x and 4x single-crosses of alfalfa. Crop Sci 34:784–788

Kidwell KK, Hartweck LM, Yandell BS, Crump PM, Brummer JE, Moutray J, Osborn TC (1999) Forage yields of alfalfa populations derived from parents selected on the basis of molecular marker diversity. Crop Sci 39:223–227

Kölliker R, Boller B, Widmer F (2005) Marker assisted polycross breeding to increase diversity and yield in perennial ryegrass (*Lolium perenne* L.). Euphytica 146:55–65

Li X, Brummer EC (2009) Inbreeding depression for fertility and biomass in advanced generations of inter- and intrasubspecific hybrids of tetraploid alfalfa. Crop Sci 49:13–19

Nei M, Li WH (1979) Mathematical model for studying genetic variation in terms of restriction endonucleases. Proc Natl Acad Sci USA 76:5269–5273

Patterson HE, Thompson R (1971) Recovery of inter-block information when block sizes are unequal. Biomerrika 58:545–554

Peakall R, Smouse PE (2006) GENALEX 6; genetic analysis in Excel. Population genetic soft ware for teaching and research. Mol Ecol Notes 6:288–295

Posselt UK (2010) Breeding methods in cross-pollinated species. In: Boller B, Posselt UK, Veronesi F (eds) Fodder crops and amenity grasses, handbook of plant breeding 5. Springer, New York, pp. 39–87

Reif JC, Melchinger AE, Xia XC, Warburton ML, Hoisington DA, Vasal SK, Srinivasan G, Bohn M, Frisch M (2003) Genetic distance based on simple sequence repeats and heterosis in tropical maize populations. Crop Sci 43:1275–1282

Robertson A, Rendel JM (1954) The performance of heifers got by artificial insemination. J Agric Sci Camb 44:184–192

Robins JG, Bushman BS, Jensen KB (2012) Dry matter yield combining ability among nine sources of orchardgrass germplasm. Euphytica 188:419–428

Tamaki H, Yoshizawa A, Fujii H, Sato K (2007) Modified synthetic varieties: a breeding method for forage crops to exploit specific combining ability. Plant Breed 126:95–100

Tamaki H, Sato K, Ashikaga K, Tanaka T, Yoshizawa A, Fujii H (2009) High-yield timothy (*Phleum pratense* L.) strains developed by 'clone and strain synthesis', a method for breeding perennial and self-incompatible crops. Grassl Sci 55:57–62

Tamaki H, Baert J, Marum P (2010) Timothy. In: Boller B, Posselt UK, Veronesi F (eds) Fodder crops and amenity grasses, handbook of plant breeding 5. Springer, New York, pp. 324–338

Tanaka T, Tamaki H, Cai H, Ashikaga K, Fujii H, Yamada T (2011a) DNA profiling of seed parents and a topcross tester and its application for yield improvement in timothy (*Phleum pratense* L.). Crop Sci 51:612–620

Tanaka T, Tamaki H, Ashikaga K, Fujii H, Yamada T (2011b) Proposal for shift to reciprocal recurrent selection in "clone and strain synthesis" timothy breeding using molecular marker diversity. Crop Sci 51:2589–2596

Tanaka T, Tamaki H, Ashikaga K, Fujii H, Sato K, Iwabuchi K, Adachi M (2013a) Evaluation of forage yield performance in advanced generations of a timothy "clone and strain synthesis" strain. Grassl Sci 59:221–225

Tanaka T, Tamaki H, Ashikaga K, Fujii H, Yamada T (2013b) Use of molecular marker diversity to increase forage yield in timothy (*Phleum pratense* L.). Plant Breed 132:144–148

Tucak M, Popovic S, Grljusic S, Cupic T, Bolaric S (2011) Implementation of molecular markers diversity in parental selection of alfalfa (*Medicago sativa* L.) germplasm. Biotechnol Biotechnol Eq 25:2310–2314

Ueda S (1990) Timothy breeding in Japan. Jpn Agric Res Q 24:195–201

Chapter 13
Single Nucleotide Polymorphism (SNP) Markers for Allele Quantification in *Lolium* (Poaceae): Development and First Applications

Tatjana Sretenovic Rajicic, Thomas Lübberstedt, Louise Bach Jensen, Uwe Scholz, W. Eberhard Weber, Andreas Graner and Klaus J. Dehmer

13.1 Introduction

Ryegrasses of the genus *Lolium* are the most important fodder and turf grasses in temperate climate regions. Two species are mostly grown: *Lolium perenne* L. (perennial ryegrass, also called English ryegrass) and *L. multiflorum* L. (annual ryegrass, also known as Italian ryegrass). Both perennial and annual grasses are wind pollinated outcrossing species, and usually form populations of highly heterozygous plants. Ryegrasses are valued for their high yield potential, fast establishment, adaptation to grazing, high palatability, and digestibility. They are also used as turf species for sports fields and recreational areas.

Changes in global climate and the necessity to maximize the use of ryegrass as fodder and turf grasses can be addressed by systematic biodiversity studies (Humphreys et al. 2005a). Genetic diversity of *Lolium* ssp. has been studied with different methods, using morphological differentiation and isozymes (Charmet and Balfourier 1994; Charmet et al. 1996; Fernando 1997; Bennett 2000; Gilliland 2000), as well as different types of DNA markers: restriction fragment length polymorphisms/RFLPs (Xu and Sleper 1994), random amplified polymorphic DNAs/RAPDs (Huff 1997; Bolaric et al. 2005), chloroplast-specific markers (Balfourier et al. 2000; McGrath et al. 2006); amplified fragment length polymorphisms/AFLPs

K. J. Dehmer (✉) · T. Sretenovic Rajicic · U. Scholz · A. Graner
Leibniz Institute of Plant Genetics and Crop Plant Research, Corrensstrasse 3, Gatersleben 06466, Germany
e-mail: dehmer@ipk-gatersleben.de

T. Lübberstedt
Agronomy Department, Iowa State University, 1204 Agronomy, Ames, IA 50011, USA

L. Bach Jensen
Illumina Denmark ApS, Ørestads Boulevard 73, 2300, København, Denmark

W. E. Weber
Institute of Agricultural and Nutritional Sciences, Plant Breeding, Martin-Luther-University Halle-Wittenberg, Halle 06099, Germany

H. Budak, G. Spangenberg (eds.), *Molecular Breeding of Forage and Turf*,
DOI 10.1007/978-3-319-08714-6_13

(Roldán-Ruiz et al. 1997; Roldán-Ruiz 2000a; b; Cresswell et al. 2001; Skøt et al. 2002; Van Treuren et al. 2005), sequence tagged sites/STS markers (Roldán-Ruiz et al. 2001; Humphreys et al. 2005b), and simple sequence repeats/SSRs (Kubik et al. 1999; Pašakinkienė et al. 2000; Jones et al. 2001; Momotaz et al. 2004; Jensen et al. 2005a; Lauvergeat et al. 2005).

Single nucleotide polymorphisms (SNPs) have recently found increasing attention in genomic research (Brookes 1999). They represent the most abundant class of DNA polymorphisms (Rafalski 2002a; Cogan et al. 2006) both in animal and plant genomes (Gupta et al. 2001). SNP markers can be developed for any gene and are locus specific, showing high data fidelity and suitability for automated high-throughput analysis (Rafalski 2002a). This allows direct analysis of sequence differences between many individuals at large numbers of loci, which make SNPs the most convenient class of polymorphism for large-scale marker analyses (Rafalski 2002b). For these reasons we developed SNP markers for the analysis of the *Lolium* collection at the genebank of the IPK Gatersleben.

The collection comprises more than 2800 accessions—mainly *L. perenne* L., *L. multiflorum* Lam., and *L. hybridum* Hausskn. In order to study this diversity and create the basis for different kinds of research on and exploitation of the IPK's *Lolium* collection, we have developed SNP markers and applied them on the entire collection. Due to the paucity of an appropriate public sequence resource for *Lolium*, we exploited the evolutionary relationships within the *Pooideae* subfamily. Comprising more than 3000 species, the *Pooideae* are rich in economically important crops, for some of which already comprehensive genomic resources are available as it is the case for barley. We have therefore decided to investigate the usefulness of barley ESTs for the development of SNP markers in ryegrass. The strategy was the following: Barley EST primer pairs were tested on five *Lolium* accessions, the obtained products were sequenced and checked for the presence of SNPs. Specific primer sets were designed and SNPs were assayed by Pyrosequencing^{TM} (PSQ). We have chosen this SNP detection platform, since it allows the quantification of allele frequencies in bulked seed samples. Compared to the estimation of allele frequencies on a single plant level, this approach can save considerable time and resources.

The present study reports on the development of a set of highly informative SNP-markers for *Lolium*. Results are presented on allele quantification (AQ) in bulked seed samples and on the usefulness of a core marker set for fingerprinting the *Lolium* collection of the IPK Genebank.

13.2 Materials and Methods

13.2.1 Plant Materials

Seeds of the *Lolium* accessions were obtained from the IPK Oil and Fodder Crop Collections at Malchow/Poel, Mecklenburg-West Pomerania. The PCR testing of barley EST-derived primer pairs was performed on five *L. perenne* plants (accessions

GR3571, GR6877, GR6900, GR6915, and GR6925). *Lolium* SNP detection was performed on a test set of 86 genebank accessions, containing *L. perenne*, *L. multiflorum*, and *L. hybridum* (Table 13.1). Developed markers were further applied on 2829 *Lolium* accessions from IPK's genebank collection, also consisting of the three mentioned species (Electronic supplement, Supplement Table 13.1, http://pgrc.ipk-gatersleben.de/lolium/).

After SNP detection in a consensus sequence derived from five single *Lolium* plants, PSQ specific primers were designed. From the genebank accessions seed bulks of three times 33 seeds per accession were used for the DNA extraction. Designed PSQ primers were used for PSQ of the DNA from the seed bulks. PSQ analysis was performed on a PSQ 96HS instrument (Qiagen, Hilden, and Germany) using the AQ mode.

For genetic mapping in *L. perenne* the VrnA mapping population of Jensen et al. 2005a was employed. The VrnA two-way pseudo-testcross mapping population consisting of 184 *Lolium perenne* F_2 genotypes was derived from a cross between a single genotype each from the Danish ecotype Falster, and the Italian variety Veyo. Two randomly selected F1 plants from this cross were used as parents of VrnA. The mapping population has previously been genotyped for linkage map construction (Jensen et al. 2005a) and analyzed for vernalization response (Jensen et al. 2005b).

13.2.2 DNA Extraction, PCR, SNP Detection and Sequencing

Leaves from single plants for the testing of barley EST primer pairs were harvested individually and genomic DNA was isolated according to Doyle and Doyle (1990). DNA from seed bulks was extracted with the use of NucleoSpin® Multi-8/96 Plant kit, according to the manufacturer's instructions.

PCR primer sequences derived from the barley ESTs were according to Stein et al. (2007). PCR was carried out in 20 µl reactions according to Kota et al. (2003), with 1.0 U of *HotStar Taq* (Qiagen). The touchdown PCR cycling profile included 15 min initial denaturation at 96°C, followed by 52 cycles with denaturation for 30 s at 95°C and extension for 1 min at 72°C. The annealing temperature was decreased in 1°C steps from 64°C in the first cycle to 56°C in the ninth cycle and then kept constant during the remaining 43 cycles, with 1 min per cycle. Final extension step was performed for 5 min at 72°C.

PCR products were purified for the sequencing reaction with the use of MinElute® 96 UF plate (Qiagen), according to manufacturer's instructions. Sequencing was performed on the MegaBACE 1000 sequencer (Amersham Biosciences Europe, Freiburg, Germany), sequencing application. Sequences were analyzed and assembly obtained with the use of Sequencher 4.5, BioEdit vs. 7.0.1 (Hall 1999) and Lasergene™ (DNASTAR, Inc, Madison, Wisconsin, USA) software packages. Putative SNPs primer pairs (Table 13.2) were designed with the use of the Assay Design software vs. 1.0.6 (Qiagen). Primers were then analyzed on seed bulks from a test set comprising 86 diverse *Lolium* accessions (Table 13.1). PCR was carried

Table 13.1 IPK accessions used in the test set with accession numbers and botanical names according to IPK passport data. *L. hybridum* = *L. boucheanum* according to (Klapp et al. 1990)

Accession	Origin	Botanical name
GR2540	The Netherlands	*Lolium multiflorum* Lam.
GR2554	France	*Lolium multiflorum* Lam.
GR2557	Japan	*Lolium multiflorum* Lam.
GR2564	Germany	*Lolium multiflorum* Lam.
GR2576	Germany	*Lolium multiflorum* Lam.
GR2607	Poland	*Lolium multiflorum* Lam.
GR2608	Romania	*Lolium multiflorum* Lam.
GR2619	Japan	*Lolium multiflorum* Lam.
GR2623	The Netherlands	*Lolium multiflorum* Lam.
GR2656	Sweden	*Lolium perenne* L.
GR2657	Poland	*Lolium perenne* L.
GR2680	The Netherlands	*Lolium perenne* L.
GR2687	The Netherlands	*Lolium perenne* L.
GR2734	USA	*Lolium perenne* L.
GR2737	Germany	*Lolium perenne* L.
GR2738	Denmark	*Lolium perenne* L.
GR2773	Hungary	*Lolium perenne* L.
GR2783	New Zealand	*Lolium perenne* L.
GR2784	Irland	*Lolium perenne* L.
GR2788	UK	*Lolium perenne* L.
GR2799	The Netherlands	*Lolium perenne* L.
GR2803	Czecheslovakia	*Lolium perenne* L.
GR2805	France	*Lolium perenne* L.
GR2935	Germany	*Lolium perenne* L.
GR2971	Germany	*Lolium perenne* L.
GR3041	Australia	*Lolium perenne* L.
GR3060	Soviet Union	*Lolium perenne* L.
GR3070	Canada	*Lolium perenne* L.
GR3073	Irland	*Lolium perenne* L.
GR3086	Italy	*Lolium perenne* L.
GR3121	Germany	*Lolium perenne* L.
GR3129	Romania	*Lolium perenne* L.
GR3136	Finland	*Lolium perenne* L.
GR3149	Soviet Union	*Lolium perenne* L.
GR3215	USA	*Lolium hybridum* Hausskn.
GR3218	Denmark	*Lolium perenne* L.
GR3236	Belgium	*Lolium perenne* L.

Table 13.1 (continued)

Accession	Origin	Botanical name
GR3287	Germany	*Lolium perenne* L.
GR3297	Germany	*Lolium perenne* L.
GR3351	Germany	*Lolium perenne* L.
GR3466	Germany	*Lolium perenne* L.
GR3560	Germany	*Lolium perenne* L.
GR3587	The Netherlands	*Lolium hybridum* Hausskn.
GR3600	Denmark	*Lolium boucheanum* Kunth.
GR3606	Denmark	*Lolium hybridum* Hausskn.
GR5092	Germany	*Lolium perenne* L.
GR5106	Germany	*Lolium perenne* L.
GR5109	Germany	*Lolium perenne* L.
GR5113	Germany	*Lolium perenne* L.
GR5432	Germany	*Lolium perenne* L.
GR5435	Iran	*Lolium perenne* L.
GR5448	Canary Island	*Lolium multiflorum* Lam.
GR5450	–	*Lolium multiflorum* Lam.
GR5452	Libya	*Lolium multiflorum* Lam.
GR5559	Romania	*Lolium perenne* L.
GR5865	Poland	*Lolium perenne* L.
GR5868	Poland	*Lolium perenne* L.
GR6032	Germany	*Lolium perenne* L.
GR6037	Germany	*Lolium perenne* L.
GR6038	Germany	*Lolium perenne* L.
GR6305	Germany	*Lolium perenne* L.
GR6338	–	*Lolium perenne* L.
GR6349	Germany	*Lolium perenne* L.
GR6392	–	*Lolium perenne* L.
GR6558	Bulgaria	*Lolium perenne* L.
GR6606	Bulgaria	*Lolium perenne* L.
GR6835	Germany	*Lolium perenne* L.
GR6865	Spain	*Lolium perenne* L.
GR6897	Spain	*Lolium perenne* L.
GR6906	Spain	*Lolium perenne* L.
GR6921	Spain	*Lolium multiflorum* Lam.
GR6938	Spain	*Lolium hybridum* Hausskn.
GR7211	Germany	*Lolium perenne* L.
GR7373	Spain	*Lolium perenne* L.
GR7677	Irland	*Lolium perenne* L.

Table 13.1 (continued)

Accession	Origin	Botanical name
GR7681	Irland	*Lolium perenne* L.
GR8401	The Netherlands	*Lolium perenne* L.
GR8421	France	*Lolium perenne* L.
GR8503	Greece	*Lolium perenne* L.
GR8508	Turkey	*Lolium perenne* L.
GR8512	Soviet Union	*Lolium perenne* L.
GR8530	Turkey	*Lolium perenne* L.
GR8818	Romania	*Lolium perenne* L.
GR9016	The Netherlands	*Lolium perenne* L.
GR9030	The Netherlands	*Lolium perenne* L.
GR9130	UK	*Lolium boucheanum* Kunth.

out with the use of the same master mix composition as previously described, with the PCR cycling profile including initial denaturation at 95°C for 15 min, followed by 50 cycles including denaturation at 95°C for 15 s, annealing at 58°C for 30 s and extension at 72°C for 30 s. Final extension was performed on at 72°C for 10 min.

Preparation for the PSQ reaction was done according to the manufacturer's instructions (Qiagen). Sequencing was done on the PSQ^{TM} HS96 A sequencing machine and the AQ protocol of the PSQ^{TM} HS96 A software (Qiagen) was applied.

13.2.3 Genetic Mapping

For genetic mapping, polymorphic SNPs in the mapping population were transformed into CAPS markers. The PCR protocol was the same as for testing barley EST primer pairs, described above. Enzymes used for the restrictions were *Bst*YI, *Hin*fI, *Ava*II, *Rsa*I, *Hpy*CH4V, *Bbv*I (New England Biolabs, Frankfurt, Germany), and the reactions were performed according to the manufacturer's instructions. In the cases where there was no restriction site present at the parental plants, the progeny was analyzed with PSQ, using the genotyping SNP mode, and SNPs were scored from the haplotypes of the parents/progeny plants.

The VrnA mapping population was originally mapped with 33 SSR and 59 AFLP markers (Jensen et al. 2005a). Briefly, map construction including the new barley EST-derived markers was carried out using the Haldane mapping function within the software JoinMap 3.0 (Van Ooijen et al. 2001). All linkage groups (LGs) were built with a minimum LOD of 4.0 except for LG1 and LG2, which grouped at LOD 5.0. The map consists of seven LGs, named according to the chromosome assignment found in the ILGI perennial ryegrass population (http://ukcrop.net/perl/ace/search/FoggDB).

Table 13.2. The list of the designed *Lolium* primer pairs and sequences that were analyzed by PSQ. Sequence to analyze: SNP codes used according to IUPAC nucleotide ambiguity codes (Cornish-Bowden 1985), SNP positions *bold* and *underlined*.

StyleMarker name	PSQ forward primer	PSQ reverse primer	PSQ sequencing primer	Sequence to analyse
GBS0031	b—GATCAGGACATCGACGAC	GCAATTATGCTCAGGTAA	r—GGTAACA-CAGAAAGCAAGAG	RGGATAGTGTCA
GBS0125	TTGAAGGATTGGGTGTTGG	b—CAAGATAC-GCCTTCTTCCG	TGGGTGTTGGGAAAG	AYACACAC TTYGAGGA ATACATAATT G
GBS0138	b—GAAGACCATCAAGAGCG-TATTGT	AGAAGGGAGAAATAGC-GACTGC	r—AGGGAGAAATAGCGACT	GCCRTTGAACAATAC-GCTCTTGATGGTCT
GBS0309	CTGGTTCATGGCAGCCACTATA	b—GCCACTTGCATCTTT-GTTACACA	TCCCATGAAGGTGGT	GAGYGAG-GCACCCATTTCAGATATCA
GBS0337	b—CGCCATCTCTCCAAATACCAA	GGCGCAATGAGCTCGTGA	r—AATGAGCTCGTGATGAG	CGASTTTATTTGTC
GBS0340	b—ACATGCAAATCAGCATCAGTT	TCCTTCTGTATCATCGA-CACTG	r—TGCTCCCATTGTGTTT	AYGAAGTCAC CG
GBS0342	TCT TGG CAA TGT GTT CAC	b—AAA CAA ATA GACGCA TCA GA	GGA AGA GGG ATG CTG	CMGCTCAGGC
GBS0467	CCATGTCCGTTTGTGCCG	b—CAAACAGCTA-CAGGGCAAACAA	AGGCGGCAACCACCAC	MATRTRGTTCGAT
GBS0489	ACGATGGACACGCTAGTATGG	b—CTACCACCAGGTGAT-CATCAGA	CTTAAGACCTGCTCTTGAA	YGGAATTGCT CTGAT-GATCA CCTGGT
GBS0505	GCCGCTGCTGCATACTGAG	b—TCTAACCACCGTCC-GCTAC	CAGGCCTTTTCCAGT	ASAGAACA
GBS0538	ATTTCGAAACGGTTGATG	b—CTGCCTGTAGAAC-CAATGAT	AACGGTTGATGACGAT	TAKGTAGTAATC
GBS0551	b—TCAACATGGAGCCTTCTTT-GTG	ACCTTAAGCTTTAGCTGT-GAACAA	r—TCCATAGCATGTGAAACT	RTTTTTCACAA AGAAGGCTCC ATGTTG
GBS0559	AAGGGGCTGGAGTTTTATGATTT	b—AGCTCGTCCAAATTC-GTGTATTTA	TGACAAATAATGATCGTTC	TGRCAACAAACGC-CACCCCAACAATACA

Table 13.2. (continued)

StyleMarker name	PSQ forward primer	PSQ reverse primer	PSQ sequencing primer	Sequence to analyse
GBS0577	AAACAAGGAAATACTC-CAAAGACA	b—ATGTGCCACAGCTTAT-GTTGTTA	AGGAAATACTCCAAAGACA	RCAGAGCATA CACCAC
GBS0594	AAAATGCCATTATTCCTCACTG	b—TGATCACCATCTTTTTG-TATCTTC	GGAGACAGATGCAGCTA	CRTCCGAGAG ATA
GBS0603	b—ATGTGCTTCCGCTCGTTCA	CGGGGAGGAGCTCAACAA	r—GAGGAGCTCAACAAGGT	YGTGGCCATGAACG
GBS0705	AGCTCTTGGAATCTTATACATT	b—GATCTTCATTCTTCTT-GCTTA	TGGAATCTTATACATTATTG	YACRAACTTGTAGAT
GBS0751	CACACTGCAAGTCCACCTCG	b—CGGATGCGGAGAGA-GAAA	CCACCTCTCGGTATGG	ATGSATGYCCCCCCTCT-TACCTCGCGCGC
GBS0766	AGAAGGAGCGGAACAGGTTAG	b—CACCGGACAGCATCA-GATT	CCTGTTGACAAAAGAACTTA	CRGTGAAAAT ATAGG-TAGTC AAAACAC
GBS0819	ATGCTTTGATGTGGCTGAGA	b—CAAAGACTGATA-ACCTCCATCACC	GTGGCTGAGAGGAGTT	RCTGGTGATG GAGGT-TATCA GTCTTT
GBS0821	b—TAATGTTGCAGCGACATGTA	GCCGGTTATTATTGTCAGC	r—ACACGTACAGAA-CAAATTAG	RTACATGTCG CTGCAA-CATT ACTACT
GBS0835	ACACAGTTCAACTCT-GCTTCTCTT	b—CGCGAGGACTTCAA-CAAAT	GTCATCTCTCACGAGC	YTATGGGAGCAACATTG-GAGGATACA

b=primer which was modified with biotin, *r*=stands for reverse sequencing primer, *bold* primer sets used in analysis of entire collection

All *Lolium* consenus sequences were analyzed with a BlastX (Altschul et al. 1990) against the NCBI NRPEP database (release 16 with 2,869,704 sequences).

13.2.4 Data Analysis

Polymorphism information content (PIC) values were calculated according to (Weber 1990; Anderson et al. 1993). Genetic distance was calculated according to (Nei 1972) with the use of NTSYSpc software (Rohlf 2002). Principal coordinate analysis (PCA) was performed in the BioNumerics software package version 4.5 (Applied Maths, Sint-Martens-Latem, Belgium). Correlations have been calculated by the use of software Mantel version 2.0.

13.3 Results

13.3.1 SNP Assay Development

We tested 264 barley EST primer pairs on five *Lolium* plants. Two hundred and fifty-one of the used ESTs are included on a transcript map of barley (Stein et al. 2007). From this number of ESTs, 97 primer pairs (37 %) were not giving any product on *Lolium* DNA. One hundred and sixty-seven barley EST primer pairs (63 %) were giving multiple or single bands after PCR amplification of *Lolium* DNA. These barley EST primers were further used in order to optimize conditions for sequencing of *Lolium* DNA. After optimizing PCR conditions, discrete amplification products were obtained from 53 barley ESTs (20 %). Upon sequencing, alignment and SNP detection, functional PSQ *Lolium* SNPs could be developed from 22 (8 %) barley EST primer pairs (Table 13.2).

Table 13.3 presents details about the developed *Lolium* SNP PSQ assays. The average number of SNPs in the 22 sequences is 7.6. Within the 22 developed assays, 26 SNPs were analyzed. The frequency of the A—G (T—C) transitions was 77 %, transversions A—C (G-T) and G—C occur at 7.6 and 11 %, respectively. All SNPs identified in this study were bi-allelic. PIC values for developed markers are presented in Table 13.3.

Lolium SNP markers were tested for polymorphism in the VrnA mapping population. Eleven markers were polymorphic and could thus be mapped (Fig. 13.1, Table 13.3). Due to the lower costs, six SNP markers were transformed into and mapped as CAPS markers, whereas five markers were mapped via PSQ (Fig. 13.1). For seven of these eleven markers the map position in barley was known (Stein et al. 2007), and in the *Lolium* genome, they were located on the homologous LGs (Table 13.3). The remaining four barley EST-derived markers (GLS0819, GLS0337, GLS0505, and GLS0603) mapped to LGs 2, 4 (twice), and 6 in ryegrass, respectively (Fig. 13.1, Table 13.3).

Table 13.3 Data on 22 *Lolium* markers. The first three columns are showing information on barley ESTs from which *Lolium* markers (column 4) are developed. Data regarding SNPs are related to *Lolium* SNPs and *Lolium* sequences. Consensus SNP count is based on the sequence analysis of the consensus sequence, provided as the electronic supplement 3. SNP position refers to the analyzed sequence, listed in Table 13.2. SNP codes are used according to IUPAC nucleotide ambiguity codes (Cornish-Bowden et al. 1985). Bolded markers were used for the analysis of the collection of 2829 accessions, the average standard deviations observed here provided in the last column (n.a.: not applicable, as marker set not employed on entire collection)

MoMa marker barley	Barley chromosome	Barley chromosome arm	*Lolium* marker	SNP code	*Lolium* linkage group	PIC values	Consensus SNP count	SNP position consensus	SNP position	SD entire collection
GBS0125	1H	Long	GLS0125	Y, Y	LG 1	0.92; 0.91	33	277, 286	2, 11	0.0425/0.0370
GBS0342	1H	Long	GLS0342	M	LG 1	0.23	3	117	2	n.a.
GBS0821	1H	Anywhere	GLS0821	R	Unknown	0.93	7	258	1	n.a.
GBS0705	2H	Long	GLS0705	Y, R	LG 2	0.89; 0.89	4	284, 287	1, 4	n.a.
GBS0766	2H	Anywhere	GLS0766	R	LG 2	0.80	8	141	2	n.a.
GBS0031	3H	Short	GLS0031	R	Unknown	0.74	2	251	1	n.a.
GBS0538	3H	Long	GLS0538	K	Unknown	0.90	4	230	3	0.0415
GBS0559	4H	Long	GLS0559	R	Unknown	0.79	1	203	3	n.a.
GBS0751	4H	Anywhere	GLS0751	S	LG 4	0.89	6	53	4	0.0673
GBS0551	4H	Anywhere	GLS0551	R	Unknown	0.83	5	303	1	n.a.
GBS0594	5H	Short	GLS0594	R	Unknown	0.40	10	74	2	n.a.
GBS0138	5H	Long	GLS0138	R	Unknown	0.72	12	311	4	0.0099
GBS0577	5H	Anywhere	GLS0577	R	Unknown	0.30	2	48	1	n.a.
GBS0489	6H	Long	GLS0489	Y	LG 6	0.85	6	112	1	0.0364
GBS0340	7H	Long	GLS0340	Y	Unknown	0.78	16	59	2	n.a.
GBS0835	7H	Anywhere	GLS0835	Y	LG 7	0.92	11	148	1	0.0463
GBS0337	unknown	–	GLS0337	S	LG 4	0.92	10	214	4	n.a.
GBS0467	unknown	–	GLS0467	M, R, R	Unknown	0.90; 0.89; 0.83	3	179, 182, 184	1, 4, 6	n.a.
GBS0603	unknown	–	GLS0603	Y	LG 6	0.87	5	194	4	n.a.
GBS0505	unknown	–	GLS0505	S	LG 4	0.89	15	115	2	n.a.
GBS0819	unknown	–	GLS0819	R	LG 2	0.92	2	419	1	0.0495
GBS0309	unknown	–	GLS0309	Y	Unknown	0.8	3	168	4	n.a.

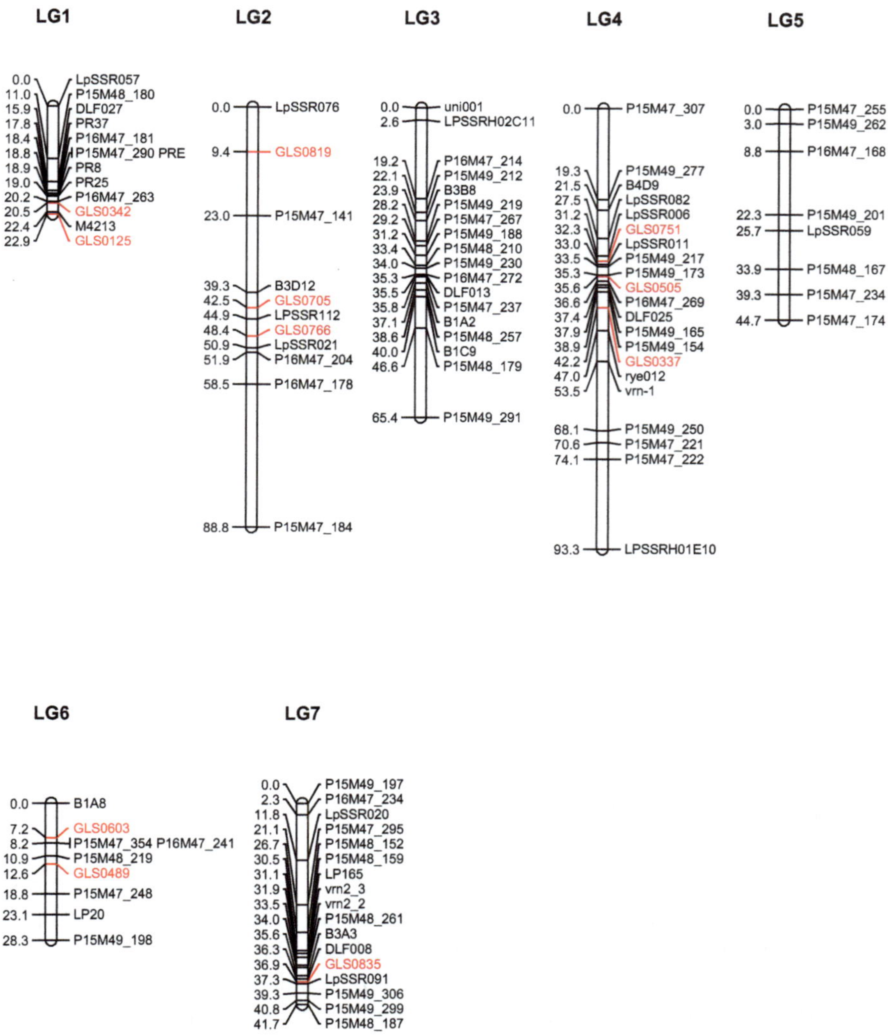

Fig. 13.1 Locations of barley EST-derived SNP markers (*red*) mapped in the *Lolium perenne* L. mapping population VrnA (Jensen et al. 2005a). GLS0125, GLS0337, GLS0489, GLS0505, GLS0705, and GLS0751 were mapped as CAPS markers. GLS0342, GLS0603, GLS0766, GLS0819, and GLS0835 were mapped as SNPs via PSQ genotyping protocol

Using the software tool "MoMaVis" (Scholz et al. unpublished; http://pgrc.ipk-gatersleben.de/lolium/momavis.php) an interactive visualization of the ryegrass maps and a comparison against the barley maps (Stein et al. 2007) is implemented. An example is shown in Fig. 13.2 and complete visualization is available in electronic supplement 2, http://pgrc.ipk-gatersleben.de/lolium/. *Lolium* consensus sequences are provided as electronic supplement 3.

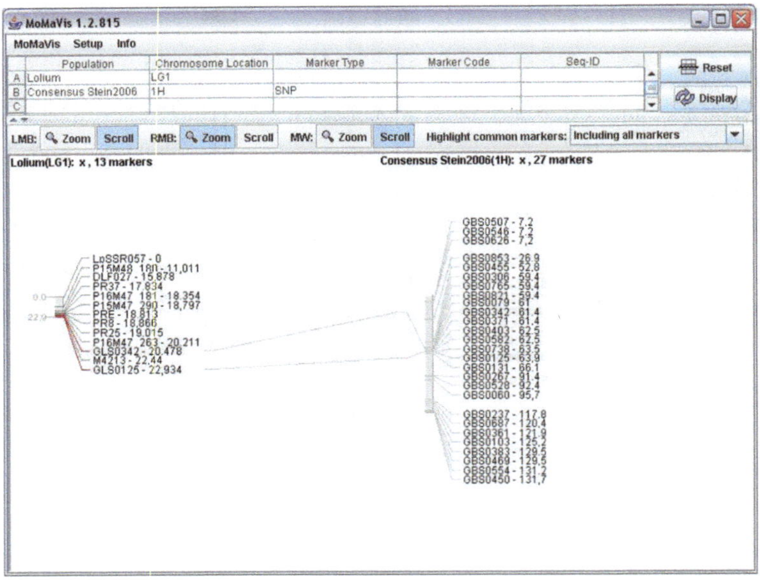

Fig. 13.2 Visualization of mapping information for *Lolium* chromosome LG1 by "MoMaVis": An example for mapping data visualisation by "MoMaVis" (Scholz et al. unpublished, http://pgrc.ipk-gatersleben.de/lolium/momavis.php) is given for the comparison of *Lolium* chromosome LG1 against all barley SNP markers mapped on chromosome 1H (Stein et al. 2007)

13.3.2 Diversity Analysis

Twenty-two SNP ryegrass markers were assayed in a test set of 86 *Lolium* accessions. DNA was extracted from the three seed bulks per accession and PSQ was performed using the AQ protocol. AQ values were loaded into the Bionumerics software and PCA analysis was performed to visualize population structure (Fig. 13.3). As expected *L. perenne* and *L. multiflorum* form two groups with the accessions of *L. hybridum* dispersed in between.

To examine the reliability of our method, we compared AQ values in several replicates. Differences between two replicas of the same assay (two PSQ runs of the same PCR) were below 1%. Differences between two technical replicates of the same assay (complete replication of the assay including two separate PCRs with the same primers) averaged at 1.16% (no data shown). By the determination of standard deviation we could collect information about its varying extents, both when analyzing SD per accession for a given primer set (Fig. 13.4) or on a per SNP primer set basis (Table 13.3).

To investigate if repeated regenerations could change the authenticity of an accession (population), individual seed lots of six randomly chosen genebank accessions from varying reproduction years and/or locations were fingerprinted. In no instance, records for a direct lineage between seed lots are present. However, this possibility might exist for at least some seed lots. The 26 respective seed lots of the

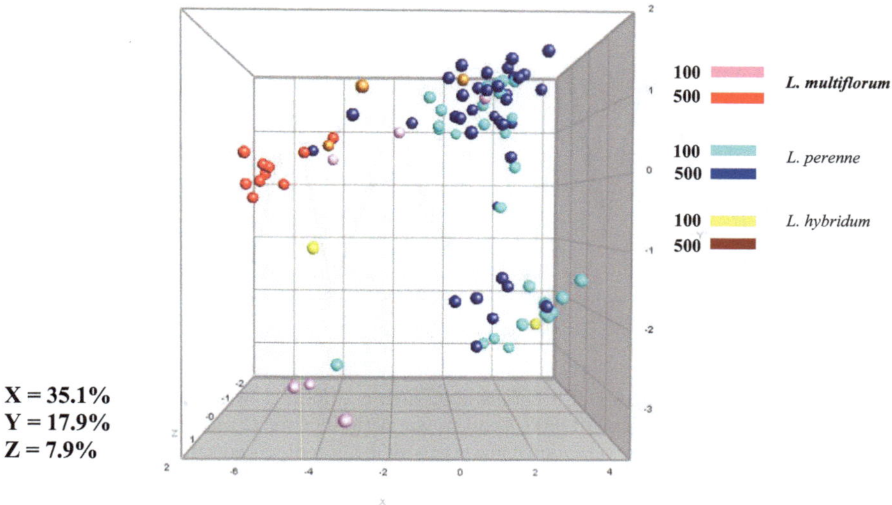

Fig. 13.3 PCA of test set of 86 *Lolium* accessions analyzed with 22 SNP markers. Status of accessions is labeled according to the IPK passport data: 100=wild 500=advanced/improved cultivar. *X*, *Y*, and *Z* represent first three principal components, explaining 35.1, 17.9, and 7.9% of the variability, respectively

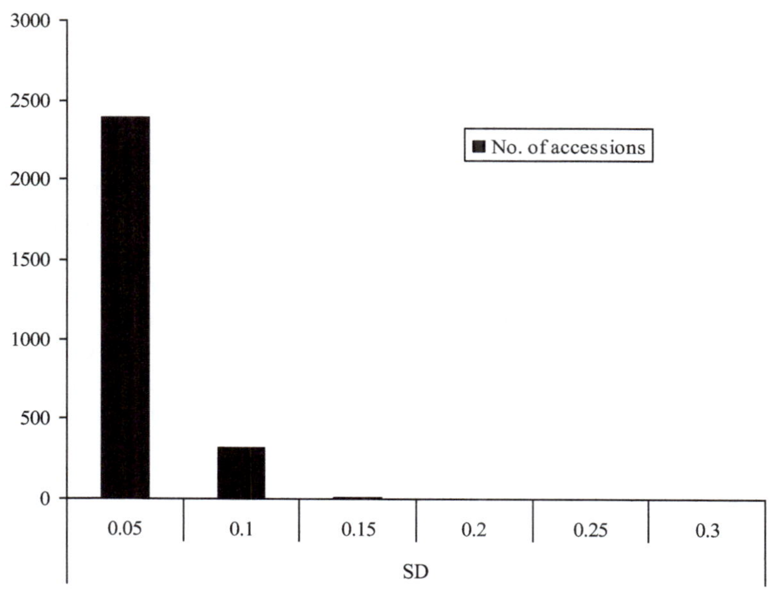

Fig. 13.4 SD values for the set of 2726 genebank accessions from the collection after employing marker GLS0125. For the most of the cases SD between the three AQ values of the three seed bulks is 0.05 or lower, in few cases it has higher values up to 0.20

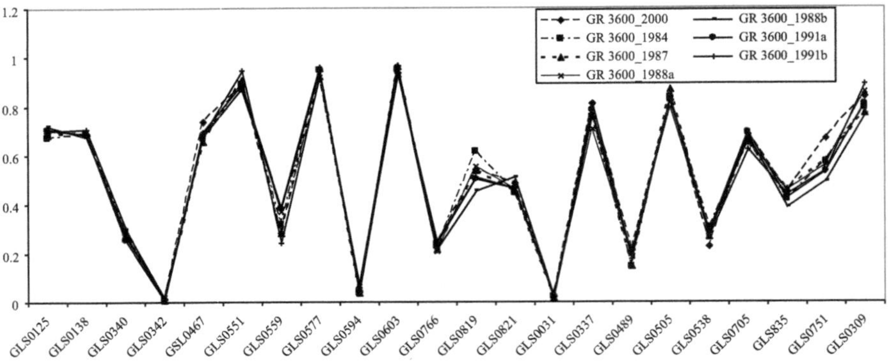

Fig. 13.5 Biological replicates of a *Lolium* accession, exemplified by the AQ values for seven different seed lots of accession GR3600. Analysis was performed for 22 markers. On the *y*-axis are AQ values. Four digits by the accession name represent the year of reproduction and *a* and *b* letters indicate multiplications in the same year

six selected accessions were analyzed with 22 markers. Cluster analysis revealed that in all cases seed lots of the same accession grouped together. However, in one case (GR2695) the similarity between one seed lot (GR2695_1988b) and the remaining three others decreased to about 98% (Fig. 13.6), which was equal to the similarity between the two accessions GR3591 and GR3600. The average SD value observed here was 0.02, these low deviations being exemplified by AQ values for 5 years of reproduction and seven seed lots for one accession (Fig. 13.5).

To develop a core set for the systematic fingerprinting in *Lolium* seven markers (GLS0125, GLS0138, GLS0489, GLS0538, GLS0751, GLS0819, and GLS0835) were selected based on their information content and an even distribution across the genome (inferred from the mapping data in *Lolium* and barley). When genetic distance matrices obtained with 7 and 22 markers were compared for test set of 86 accessions, correlation coefficient was 0.92, at the $p=0.005$. PIC values ranged from 0.72 to 0.92 (Table 13.3). Assuming a high degree of synteny between the *Lolium* LGs displayed in Fig. 13.1 and the barley transcript map of (Stein et al. 2007), markers should be located on at least six of the seven chromosomes (marker GLS0819 remained unassigned). Fingerprint analysis of the entire collection and subsequent PCA analysis again confirms the distinction between the species *L. perenne* and *L. multiflorum* (Fig. 13.7) and the intermediate position of *L. hybridum*.

13.4 Discussion

13.4.1 Development of Lolium SNP Markers

The rationale for the development of *Lolium* SNP markers from barley ESTs has been based on the relatively close taxonomic relationship of the genus *Lolium* to

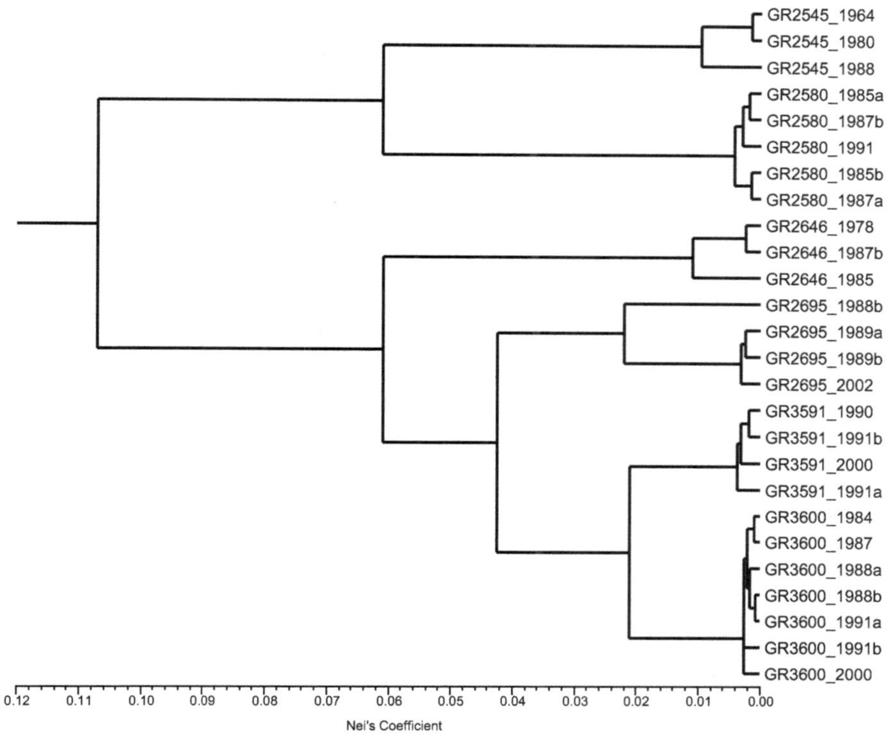

Fig. 13.6 Comparison of different seed lots from six *Lolium* accessions. The last four digits represent the year of seed production, with *a* and *b* indicating different seed lots generated independently, but in the same year

the Triticeae (e.g., *Triticum*, *Hordeum*, and *Secale*), which reside within the same subfamily *Pooideae* and the high level of synteny (i.e., marker colinearity) present within the *Poaceae* family (Devos and Gale 1997; Jones et al. 2002).

There are several reports showing the use of ESTs for development of different marker sets transferable across related plant species, based on the synteny presumptions. Lem and Lallemand (2003) found that about 68 % of the STS markers derived from different Gramineae species showed length polymorphisms in *Lolium*. Thiel et al. (2003) reported that about 60 % of the barley EST-SSR primers yielded amplification in wheat and rye, while in rice this rate was still 40 %. Taylor et al. (2001) were using primers derived from three barley genes and a gene from the Mediterranean grass *Phalaris coerulescens* to amplify homologous fragments in *L. perenne*, and reported that 52 % primer pairs amplified homologous sequences in *L. perenne*. In our case 63 % of barley EST-derived primer pairs amplified *Lolium* sequences, and of the corresponding 167 primer pairs, 53 (32 %) were useful for the generation of *Lolium* sequences. From this number, 22 barley EST primer pairs amplified polymorphic sequences useful for, which is 13 % of the starting number of barley EST primer pairs that amplified in *Lolium* (167) and 8 % of the total number of applied barley ESTs (264). The reduction from 53 amplification products to 22 functional

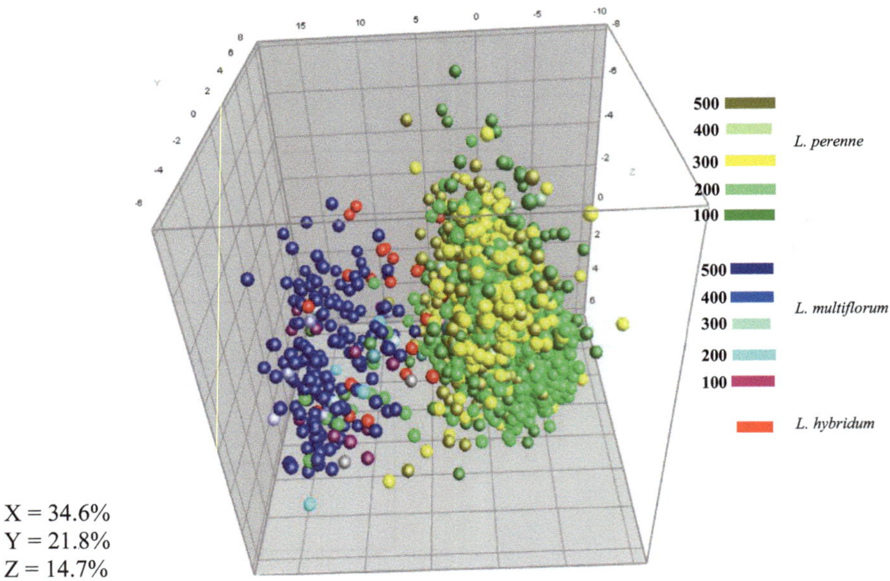

Fig. 13.7 PCA of the 2829 accessions from the *Lolium* collection analyzed with seven markers (GLS0125, GLS0138, GLS0489, GLS0538, GLS0751, GLS0819, and GLS0835). 500 = advanced/improved cultivar, 400 = breeding/research material, 300 = traditional cultivar/landrace, 200 = weedy, 100 = wild

Lolium markers was caused by the following factors: (i) incomplete sequence sets (consensus from at least four out of five plants could not be obtained) were dropped from further analysis; (ii) monomorphic sequences, or (iii) the SNP position did not permit a proper placement of PSQ primers. From 22 developed assays (26 SNPs), 11 could be genetically located in the mapping population used. Marker pairs developed from barley ESTs with known mapping positions were mapping to homologous LGs in ryegrass. This is in-line with previous reports (Jones et al. 2002; Alm et al. 2003; Van Deynze et al. 1998). For four markers (GLS0819, GLS0337, GLS0505, and GLS0603) the map position in barley is not known so far. Based on the map positions in ryegrass and the high colinearity between some Triticeae and ryegrass (Alm et al. 2003), the respective ESTs can be expected to map to barley chromosomes 2, 4, and 6, respectively. However, it should be noted in this context that chromosomes 4 and 6 are not 100% syntenic between barley and ryegrass (Alm et al. 2003).

13.4.2 Application of Barley Derived SNP Markers in Ryegrass

Two major groups were observed when employing the above mentioned SNP markers, consisting of *L. perenne* and *L. multiflorum* accessions. As expected, *L. hybridum* accessions were dispersed between the two species in PCA, since this species is a hybrid between *L. perenne* and *L. multiflorum*. Accessions were also grouping into wild material and cultivars. However, this distinction was not complete, especially within *L. perenne*. Within *L. multiflorum* accessions, cultivars were grouping together, while wild-type accessions showed higher levels of diversity.

It is known that *Lolium* accessions are frequently difficult to distinguish, since different *Lolium* species can hybridize readily (Charmet et al. 1996). With other grass genera, like *Festuca*, hybridization has been reported (Momotaz et al. 2004). For ryegrass, variety distinction is also difficult (Roldán-Ruiz et al. 2001; Lem and Lallemand 2003). Both our tests of markers on a smaller sample (test set of 86 accessions with 22 markers) and on the entire collection (7 markers) are confirming the existence of 2-tiered structure within the *Lolium* collection. Even with only seven markers employed on the whole collection we were able to distinguish between the two species *L. perenne* and *L. multiflorum*.

In our approach, we were not genotyping material for the presence of certain alleles within detected SNPs, in which usually two and at maximum four alleles can be detected. Instead, we were fingerprinting our accessions by quantification of allele frequencies within the bulk of each accession. This approach has created quite different data than simple (0–1) genotyping would do—each marker can obtain 0–100 % frequency within investigated bulks. This means that for one SNP we have theoretically (if the difference in AQs between two measurements is at 1–2 %), 50–100 data points, 100 "polymorphisms." Even if we take into account the technical accuracy of the PSQ method, which was reported to detect allele frequency differences of less than 2 % (Gruber et al. 2002) we could still generate enough data points to distinguish among accessions with seven markers employed. Of course, not all of the SNPs are having the same PIC value, extents of standard deviation and not all of them are "covering" the complete range of 0–100 %, but for the selected core set of seven markers, the information content is very high. This is also evidenced by the differentiation of *L. perenne* and *L. multiflorium*, which was obtained by using all 22 SNP markers on the test set, but also by the core marker set only. Correlation between distance matrices obtained with 7 and 22 markers is high and significant.

Over-all, the employment of the AQ method via PSQ for fingerprinting has proven to be a reliable and robust technique. In their studies of PSQ-based SNP allele frequency in DNA pools, Lavebratt et al. (2003) are reporting a technical precision of 1.73 % for the method. In our findings collected from the test set of accessions, variation among the replicates of two different PCR products—technical precision of the method—was on average 1.63 %. Standard deviations among biological replicates were 2 % on average for the multiple seed lots per accession and 4 % for the entire genebank collection. The extent of the standard deviation proved to be dependant on the SNP primer set being employed. Here, differences of almost a factor of seven do exist (GLS0138 vs. GLS0751, cf. Table 13.3), which might be reducible by further optimizations in PCR conditions of the primers of the respective primer set.

13.4.3 Toward a Marker Assisted Collection Management

Conservation of wind pollinated accessions is difficult. Maintenance of such accessions in the genebank demands repeated regenerations of samples, and it is very important to gain information about genetic integrity of the propagated accessions. Differences in allele frequencies after a few years of reproduction of outcrossing populations can occur due to few factors, like selection or drift (Chebotar et al. 2003; Van Hintum 2007). For open-pollinated species, each regeneration might result in a change of allele frequency of the accession as the consequence. Chebotar et al. (2003) showed that in rye accessions fingerprinted with microsatellites, differences in allele frequencies between obtained subpopulations after one or two multiplication cycles were not significant for the most of the loci tested. Information about the changes in allele frequency after 1 year of reproduction and its comparison with the existing data is important for the conclusions about the genetic identity of the accession. With the markers developed, this task is rather easy for the *Lolium* accessions. Our data show that after few reproduction cycles there are slight differences in allele frequencies among the subpopulations of the same accession, but the accessions are still unique, i.e., can be differentiated from each other. This provides circumstantial evidence that the management of the collection is sufficient to preserve the authenticity of an accession even after several rounds of multiplication.

Acknowledgments The authors would like to thank Dr. Julia Mattner for the involvement in the establishing phase, Evelin Willner for providing the seed samples plus information on them, Jacqueline Pohl and Ulrike Beier for technical support, Dr. Frank Blattner (IPK Gatersleben) for fruitful discussions, Dr. AnnaLotta Schiller and Dr. Michael Utting (Biotage AB/Qiagen) for help in assays development. This work was financed by the BMBF (Project reference number 0312830; project title: Aufbau einer bundeszentralen *Ex-situ*-Genbank für landwirtschaftliche und gartenbauliche Kulturpflanzen: Zusammenführung der Genbanken des IPK und der BAZ Braunschweig).

References

Alm V, Fang C, Busso CS, Devos KM, Vollan K, Grieg Z, Rognli A (2003) A linkage map of meadow fescue (*Festuca pratensis* Huds.) and comparative mapping with other Poaceae species. Theor Appl Genet 108:25–40

Altschul SF, Gish W, Miller W, Myers EW, Lipman DJ (1990) Basic local alignment search tool. J Mol Biol 215:403–410

Anderson JA, Churchill GA, Autrique JE, Tanksley SD, Sorrells ME (1993) Optimizing parental selection for genetic linkage maps. Genome 36:181–186

Balfourier F, Imbert C, Charmet G (2000) Evidence for phylogeographic structure in *Lolium* species related to the spread of agriculture in Europe. A cpDNA study. Theor Appl Genet 101:131–138

Bennett SJ, Hayward MD, Marshall DF (2000) Morphological differentiation in four species of the genus *Lolium*. Genet Resour Crop Evol 47:247–255

Bolaric S, Barth S, Melchinger AE, Posselt UK (2005) Molecular genetic diversity within and among German ecotypes in comparison to European perennial ryegrass cultivars. Plant Breed 124:257–262

Brookes AJ (1999) The essence of SNPs. Gene 234:177–186

Charmet G, Balfourier F (1994) Izozyme variation and species relationships in the genus *Lolium* L. (ryegrasses, Graminaceae). Theor Appl Genet 87:641–649

Charmet G, Balfourier F, Chatard V (1996) Taxonimic relationships and interspecific hybridization in the genus *Lolium* (grasses). Genet Resour Crop Evol 43:319–327

Chebotar S, Roeder MS, Korzun V, Saal B, Weber WE, Boerner A (2003) Molecular studies on genetic integrity of open-pollinated species rye (*Secale cereale* L.) after long-term genebank maintenance. Theor Appl Genet 107:1469–1476

Cogan NOI, Ponting RC, Vecchies AC, Drayton MC, George J, Dracatos PM, Dobrowolski MP, Sawbridge TI, Smith KF, Spangenberg GC, Forster JW (2006) Gene-associated single nucleotide polymorphism discovery in perennial ryegrass (*Lolium perenne* L.). Mol Genet Genomics 276:101–112

Cornish-Bowden A (1985) IUPAC-IUB symbols for nucleotide nomenclature. Nucleic Acids Res 13:3021–3030

Cresswell A, Sackville Hamilton NR, Roy K, Viegas BMF (2001) Use of fragment length polymorphism markers to assess genetic diversity of *Lolium* species in Portugal. Mol Ecol 10:229–241

Devos KM, Gale MD (1997) Comparative genetics in the grasses. Plant Mol Biol 35:3–15

Doyle JJ, Doyle JL (1990) Isolation of plant DNA from fresh tissue. Focus 12:13–15

Fernando WMU, Hayward MD, Kearsey MJ (1997) Isozyme and quantitative traits polymorphisms in European provenances of perennial ryegrass (*Lolium perenne* L). Euphytica 93:263–269

Gilliland TJ, Coll R, Calsyn E, De Loose M, Van Eijk MJT, Roldan-Ruiz I (2000) Estimating genetic conformity between related ryegrass (*Lolium*) varieties. 1. Morphology and biochemical characterisation. Mol Breed 6:569–580

Gruber JD, Colligan PB, Wolford JK (2002) Estimation of single nucleotide polymorphism allele frequency in DNA pools by using pyrosequencing. Hum Genet 110:395–401

Gupta PK, Roy JK, Prasad M (2001) Single nucleotide polymorphisms: a new paradigm for molecular marker technology and DNA polymorphism detection with emphasis on their use in plants. Curr Sci India 80:524–535

Hall TA (1999) BioEdit: a user friendly biological sequence alignment editor and analysis program for Windows95/98/NT. Nucl Acids Symp Ser 41:95–98

Huff DR (1997) RAPD characterisation heterogeneous perennial ryegrass cultivars. Crop Sci 37:557–564

Humphreys MW, Yadav RS, Cairns AJ, Turner LB, Humphreys J, Skøt L (2005a) A changing climate for grassland research. New Phytol 169:9–26

Humphreys J, Harpner JA, Armstead IP, Humphreys MW (2005b) Introgression-mapping of genes for drought resistance transferred from *Festuca arundinacea* var. *glaucescens* into *Lolium multiflorum*. Theor Appl Genet 110:579–587

Jensen LB, Aarens P, Andersen CH, Holm PB, Ghesquiere M, Julier B, Lübberstedt T, Muylle H, Nielsen KK, De Riek J, Roldán-Ruiz I, Roulund N, Taylor C, Vosman B, Barre P (2005a) Development and mapping of a public reference set of SSR markers in *Lolium perenne* L. Mol Ecol Notes 5:551–557

Jensen LB, Andersen J, Frei U, Xing Y, Taylor C, Holm P, Lübberstedt T (2005b) QTL mapping of vernalization response in perennial ryegrass reveals cosegregation with an orthologue of wheat *VRN1*. Theor Appl Genet 110:526–537

Jones ES, Dupal MP, Kölliker R, Drayton MC, Forster JW (2001) Development and characterisation of simple sequence repeat (SSR) markers for perennial ryegrass (*Lolium perenne* L.). Theor Appl Genet 102:405–415

Jones ES, Mahoney NL, Hayward MD, Armstead IP, Gilbert Jones J, Humphreys MO, King IP, Kishida T, Yamada T, Balfourier F, Charmet G, Forster JW (2002) An enhanced molecular marker based genetic map of perennial ryegrass (*Lolium perenne*) reveals comparative relationships with other Poaceae genomes. Genome 45:282–295

Klapp E, Opitz von Boberfeld W (1990) Taschenbuch der Gräser. Erkennung und Bestimmung, Standort und Vergesellschaftung, Bewertung und Verwendung. 12th revised edition. Verlag Paul Parey, Berlin, pp. 282

Kota R, Rudd S, Facius A, Kolesov G, Thiel T, Zhang H, Stein N, Mayer K, Graner A (2003) Snipping polymorphisms from large EST collections in barley (*Hordeum vulgare* L.). Mol Genet Genomics 270:24–33

Kubik C, Meyer WA, Gaut BS (1999) Assessing the abundance and polymorphism of simple sequence repeats in perennial ryegrass. Crop Sci 39:1136–1141

Lauvergeat V, Barre P, Bonnet M, Ghesquiere M (2005) Sixty sample sequence repeat markers for use in the *Festuca—Lolium* complex of grasses. Mol Ecol Notes 5:401–405

Lavebratt C, Sengul S, Jansson M, Schalling M (2003) Pyrosequencing™-based SNP allele frequency estimation in DNA pools. Hum Mutat 23:92–97

Lem P, Lallemand J (2003) Grass consensus STS markers: an efficient approach for detecting polymorphism in *Lolium*. Theor Appl Genet 107:1113–1122

McGrath S, Hodkinson TR, Salamin N, Barth S. (2006) Development and testing of novel chloroplast microsatellite markers for *Lolium perenne* and other grasses (Poaceae) from *de novo* sequencing and in silico sequences. Mol Ecol Notes 6:449–452

Momotaz A, Forster JW, Yamada T (2004) Identification of cultivars and accessions of *Lolium*, *Festuca* and *Fesulolium* hybrids through the detection of simple sequence repeat polymorphism. Plant Breed 123:370–376

Nei M (1972) Genetic distance between populations. Am Nat 106:283–292

Pašakinkienė I, Griffiths CM, Bettany AJE, Paplauskienė V, Humphreys MW (2000) Anchored simple-sequence repeats as primers to generate species-specific DNA markers in *Lolium* and *Festuca* grasses. Theor Appl Genet 100:384–390

Rafalski JA (2002a) Application of single nucleotide polymorphisms in crop genetics. Curr Opin Plant Biol 5:94–100

Rafalski JA (2002b) Novel genetic mapping tools in plants: SNPs and LD-based approaches. Plant Sci 162:329–333

Rohlf FJ (2002) NTSYSpc: numerical taxonomy system, ver. 2.11a. Exeter Software, Setauket

Roldán-Ruiz I, Van Laecke K, De Riek J, Calsyn E, DepickerA, Van Bockstaele E, Beerepoot L, De Loose M (1997) The use of molecular markers for assessing genetic conformity between varieties of ryegrass. Med Fac Landbouw Univ Gent 62/4a:1451–1457

Roldán-Ruiz I, Dendauw J, Van Bockstaele E, Depicker A, De Loose M (2000a) AFLP markers reveal high polymorphic rates inryegrasses (*Lolium* spp.). Mol Breed 6:125–134

Roldán-Ruiz I, Calsyn E, Gilliland TJ, Coll R, van Eijk MJT, De Loose M (2000b) Estimating genetic conformity between related ryegrass (*Lolium*) varieties. 2. AFLP characterization. Mol Breed 6:593–602

Roldán-Ruiz I, Van Eeuwijk FA, Gilliland TJ, Dubreuil P, Dillman C, Lallemand J, De Loose M, Baril CP (2001) A comparative study of molecular and morphological methods of describing relationships between perennial ryegrass (*Lolium perenne* L.) varieties. Theor Appl Genet 103:1138–1150

Skøt L, Sackville Hamilton NR, Mizen S, Chorlton KH, Thomas ID (2002) Molecular genecology of temperature response in *Lolium perenne*: 2. association of AFLP markers with ecogeography. Mol Ecol 11:1865–1876

Stein N, Prasad M, Scholz U, Thiel T, Zhang H, Wolf M, Kota R, Varshney RK, Perovic D, Grosse I, Graner A (2007) A 1000 loci transcript map reveals insight into barley genome evolution. Theor Appl Genet 114:823–839

Taylor C, Madsen K, Borg S, Møller MG, Boelt B, Holm PB (2001) The development of sequence-tagged sites (STSs) in *Lolium perenne* L.: the application of primer sets derived from other genera. Theor Appl Genet 103:648–658

Thiel T, Michalek W, Varshney RK, Graner A (2003) Exploiting EST databases for the development and characterization of gene-derived SSR-markers in barley (*Hordeum vulgare* L.). Theor Appl Genet 106:411–422

Van Deynze AE, Sorrels ME, Park WD, Ayres NM, Fu H, Cartinhour SW, Paul E, McCouch SR (1998) Anchor probes for comparative mapping of grass genera. Theor Appl Genet 97:356–369

Van Hintum ThJL, Van de Wiel CCM, Visser DL, Van Treuren R, Vosman B (2007) The distribution of genetic diversity in a *Brassica oleareacea* gene bank collection related to the effects on diversity of regeneration, as measured with AFLPs. Theor Appl Genet 114:777–786

Van Ooijen JW, Voorrips RE (2001) Joinmap 3.0, Software for the calculation of genetic linkage maps. Plant Research International, Wageningen, The Netherlands

Van Treuren R, Bas N, Goossens J, Jansen J, Van Soest LJM (2005) Genetic diversity in perennial ryegrass and white clover among old Dutch grasslands as compared to cultivars and nature reserves. Mol Ecol 14:39–52

Weber JL (1990) Informativeness of the human (dC-dA)*n*. (dG-dT)*n* polymorphisms. Genomics 7:524–530

Xu WW, Sleper DA (1994) Phylogeny of tall fescue and related species using RFLPs. Theor Appl Genet 88:685–690

Chapter 14
Origins of Diploid *Dactylis* from China as Determined by ITS Sequences

A. V. Stewart and N. W. Ellison

14.1 Introduction

Dactylis glomerata has a wide natural distribution over much of temperate Europe, Asia, North Africa, and the Canary Islands. Tetraploids of this species are distributed over almost the entire area while diploids have a very limited range with at least 21 disjunct populations known. At least one unique hexaploid population is also known to occur in northern Egypt (Jones 1962).

It is clear that the polyploids have developed from diploid and the likely reason why the tetraploids have such a broad distribution is that they have been much more successful in recolonizing northern regions after the last glaciation event. The distribution of the diploids is very restricted and consistent with known glacial refugia, with one exception. The subspecies *aschersonia* occurs within the Caucasus glacial refuge and has also succeeded in expanding north over a large part of Europe sympatric with tetraploids forms. It now appears likely that the success of the tetraploids was due to an enhanced vigor as the result of sexual polyploidization via 2n gametes between various diploid populations. Tetraploids have more polymorphic loci than diploids (0.80 compared to 0.70), higher heterozygosity (0.43 compared to 0.17), and a greater number of alleles per locus (2.36 compared to 1.51) (Soltis and Soltis 1993). The hybrid vigor obtained from these interecotypic tetraploids readily allowed them to dominate the postglacial expansion to such an extent that we have an almost continuous distribution of tetraploids over the full range of the species today (Lumaret 1986, 1988).

A. V. Stewart (✉)
PGG Wrightson Seeds, PO Box 175, Lincoln 7640, New Zealand
e-mail: astewart@pggwrightsonseeds.co.nz

N. W. Ellison
AgResearch Grasslands, Private Bag 11008, Palmerston North, New Zealand

H. Budak, G. Spangenberg (eds.), *Molecular Breeding of Forage and Turf*,
DOI 10.1007/978-3-319-08714-6_14

The diploid *Dactylis glomerata* subsp. *sinensis* Camus occurs in cool temperate forest zones of Sichuan, Hubei, Guizhou, Xinjiang, and Yunnan provinces in central China, at altitudes of 1000–3800 m. This distribution is disconnected from the larger contiguous zone of all remaining diploid and tetraploid *Dactylis glomerata* (Stebbins and Zohary 1959). Although originally thought to be a disjunct extension of adjacent diploid *himalayensis* of the Himalayas (Stebbins and Zohary 1959; Borrill 1977), the isozyme pattern of *sinensis* is quite different from *himalayensis* (Lumaret 1987) suggesting a separate origin. Preliminary molecular studies of a single diploid Chinese population suggested that it exhibits its greatest molecular similarity to European *aschersoniana* rather than to *himalayensis* (Stewart and Ellison 2010). We have accessed two diploid populations and ten tetraploid populations to enable this hypothesis to be more fully tested.

14.2 Methods

Twelve accessions of *Dactylis glomerata* that were believed to be original Chinese germplasm rather than introduced commercial material were obtained from various sources in China. Flow cytometry was performed to determine ploidy level with ten confirmed as tetraploids and two as diploid *Dactylis glomerata* subsp. *sinensis*. For each accession of each subspecies, three or more genotypes were sampled. The overwhelming majority of commercial seed traded internationally of *Dactylis glomerata* are tetraploid and we can therefore be reasonably certain that the two diploid accessions are of original Chinese origin but it is possible some of the tetraploids are of introduced origin. All these accessions were compared with the limited range of accessions previously obtained and published by Stewart and Ellison (2011).

The samples were studied using the sequences of the trnL (UAA) gene intron of chloroplast DNA (cpDNA) and the internal transcribed spacer regions ITS1 and ITS2 of nuclear ribosomal DNA. DNA was extracted from approximately 10 mg of dried leaf material using the DNeasy Plant Mini Kit (QIAGEN) according to the manufacturer's instructions. The ITS region was amplified using the primers EC-1 and EC-2 (Williams et al. 2001), and the trnL (UAA) intron was amplified using the primers "c" and "d" (Taberlet et al. 1991). PCR products were purified and sequenced directly as described in Ellison et al. (2006). DNA sequences were aligned, with manual adjustment to optimize alignments where necessary, using MegAlign (DNASTAR). The molecular techniques have shown their power to explain the history but only when appropriate populations are available for study (Sahuquillo and Lumaret 1999; Stewart and Ellison 2011; Stewart et al. 2013).

14.3 Results

The results of the ITS sequencing for a range of diploid forms is presented in Table 14.1. There was clear differentiation between three major groups; a European and North African clade (*aschersoniana, woronowii, smithii, santai, castellata* and

Table 14.1 Molecular ITS nucleotide differences among *Dactylis* accessions

–	–	510	613	90	92	571	601	55	425	105	562	660	475	592	649	610	627	190
Lamarkia aurea	–	G	C	C	C	A	T	C	A	C	C	A	G	–	C	G	G	T
Parthiana	K1846	G	G	T	G	A	A	C	A	C	C	A	G	–	C	G	C	C
Judacea	K1867	G	G	T	G	A	A	C	A	C	C	A	G	–	C	G	C	T
Himalayenis	K2003, K2365	G	G	T	G	A	A	C	A	C	C	A	T	–	C	G	C	T
Chinese 4x	Cultivar boaxing	T	A	C	C	G	G	T	A	C	C	A	G	–	C	G	C	T
Chinese 4x	Cultivar qiancao No 4	T	A	C	C	G	G	T	A	C	C	–	G	–	C	G	C	T
Chinese 4x	Xinjiang K3364	T	A	C	C	G	G	C/T	A	C	C	A	G	–	C	G	C	T
Chinese 4x	Sichuan K3363	T	A	C	C	G	G	C	A	T	C	A	G	–	C	G	C	T
Chinese 4x	Cultivar Chuan Dong	T	A	C/T	C	G	G	T	A	C	C	A	G	–	C	G	C	T
Chinese 4x	Guizhou K3362	T	A	C/T	C	G	G	T	A	C	C	A	G	–	C	G	C	T
Chinese 4x	Cultivar gulin	T	A	C/T	C	G	G	C/T	A	C	C	A	G	–	C	G	C	T
Chinese tetraploid	Yunnan K3365	T	A	C/T	C	G	A/G	T	A	C	C	A	G	–	C	G	C	T
Chinese 4x	Yunnan 02–106	T	A	C	C	G	G	T	A	C	C	A/-	G	–	C	G	C	T
Chinese 4x	Sichuan 02–115	T	A	C	C	G	G	T	A	C	C	A/-	G	–	C	G	C	T
Chinese 2x	Boaxing, sichuan 02–107	T	A	C	C	G	G	C	A	T	T	–	G	–	C	G	C	T
Chinese 2x	Sichuan JA07–1	T	A	C	C	G	G	C	A	T	T	–	G	–	C	G	C	T
Aschersoniana	K2306	T	A	C	C	G	G	T	A	C	C	A	G	–	C	G	C	T
Woronowii	K2341, K2663	T	A	C	C	G	G	T	A	C	C	A	G	–	C	G	C	T
Smithii	K2367	T	A	C	C	G	G	T	A	C	C	A	G	–	C	G	C	T
Santai	K2322	T	A	C	C	G	G	T	A	C	C	A	G	–	C	G	C	T

Table 14.1 (continued)

–	–	510	613	90	92	571	601	55	425	105	562	660	475	592	649	610	627	190
Castellata	K2018	T	A	C	C	G	G	T	A	C	C	A	G	–	C	G	C	T
Mairei	K2319	T	A	C	C	G	G	T	A	C	C	A	G	–	C	A	T	T
Ibizensis	PI237609	G	A	C	G	A	A	C	G	C	C	A	G	T	C	G	C	T
Lusitanica	K2216, K2592	T	A	C	G	A	A	C	G	C	C	A	G	–	C	G	C	T
Juncinella	K2217	T	A	C	G	A	A	C	G	C	C	A	G	–	A	G	C	T
Izcoi	K2326, K2348, K2593	T	A	C	G	A	A	C	G	C	C	A	G	–	A	G	C	T

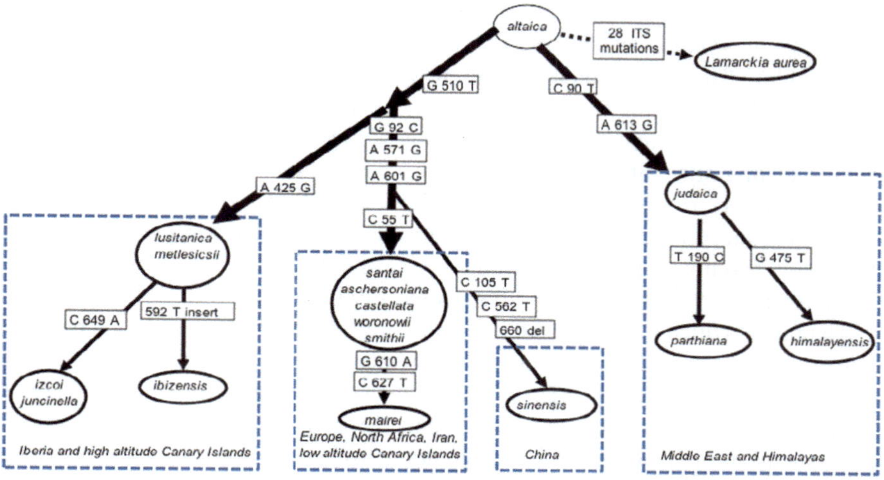

Fig. 14.1 Molecular descent in diploid *Dactylis glomerata* using ITS sequences(Numbers refer to nucleotide mutations in the ITS sequence, e.g., C 90 T is the change from C to T at nucleotide sequence position 90)

mairei); an Iberian clade (*ibizensis, lusitanica, juncinella* and *izcoi*), and a third clade of Middle-eastern (*judaica*), high altitude Iranian (*parthiana*) and the Himalayan diploids (*himalayensis*).

The Chinese diploids show similarity to the European forms but with four additional mutations. Five of the ten Chinese tetraploids express some nucleotide ambiguity at sequence positions suggesting they are hybrids of a European *aschersoniana* with the diploid Chinese forms. Four more Chinese tetraploids express some of the nucleotide ambiguity expected of a hybrid between the Himalayan *himalayensis* and the diploid European *aschersoniana* forms. The remaining tetraploid accession has a typical *aschersoniana* sequence suggesting it has developed from sexual polyploidization of two diploid *aschersoniana* parents via 2n eggs and 2n gametes.

The ancestral sequence reconstruction using the methods of Blanchette et al. (2008) for the molecular ITS sequence is outlined in Fig. 14.1.

14.4 Discussion

The results for the two Chinese diploids show an affinity with the European group (Table 14.1) although they have additional nucleotide mutations. This suggests that they have developed from an ancestral "pre-European" type population which has migrated to China at some stage as per Fig. 14.1. The nearest that pre-European *aschersoniana* type material comes to China is in the former glacial refuge of the Caucasus Mountains.

Five of the ten tetraploids have nucleotide ambiguity typical of an expected hybrid of *aschersoniana* and *himalayensis* and this is suggestive that *aschersoniana* migrated into China via a route close to the Himalayas.

Four of the remaining six tetraploids have nucleotide ambiguity typical of a hybrid between the diploid Chinese *sinensis* and *aschersoniana.* This is again suggestive of migration of *aschersoniana* into China, although in this case it could represent a very recent hybridization of Chinese diploids with commercial tetraploids introduced from Europe. The prevalence of 2n gametes in *Dactylis* ensures that there is continual potential for ongoing sexual polyploidization in this species (Lumaret et al. 1989; Lumaret and Barrientos 1990; Lumerat et al. 1992).

The rate at which ITS sequence mutations are fixed within a population depends on many factors apart from inherent mutation rate and this provides difficulty in calibrating a "molecular clock." These factors include the generation interval created by annual or perennial habit, differences between self- and cross-pollinated breeding systems and any bottlenecks in population size. In *Dactylis,* we have observed no more than three to seven ITS mutations in any of the diploid lineages and we have observed very few mutations in populations that have migrated from glacial refugia. In the annual grasses barley and maize, the ITS rate has been calculated to be approximately one mutation every 23,000 years (Zurawski et al. 1984), while the rate in perennial grasses should be slower and potentially similar to that observed in perennial *Phleum*, where Stewart et al. (2009) observed the most rapid lineage in *Phleum* to be one mutation in every 30,000 years. If this rate is similar in *Dactylis* then this genus has a relatively recent evolutionary origin and probably only arose in the intergalacial period 150,000 years ago.

The question still remains as to whether the migration into China of diploid *Dactylis* was preglacial or was it a recent post glacial migration, perhaps an example of early human transport along the Silk Road as has been shown to have occurred with white flowered Chinese kale *Brassica oleracea* (Maly et al. 1987; Gladis and Hammer 2003).

References

Blanchette M, Diallo AB, Green ED, Miller W, Haussler D (2008) Computational reconstruction of ancestral DNA sequences. Methods in molecular biology. In: Murphy WJ (ed) Phylogenomics. Humana Press, Totowa, pp 171–184

Borrill M (1977) Evolution and genetic resources in Cocksfoot. Annu Rep Welsh Plant Breed Stat, pp 90–209

Ellison NW, Liston A, Steiner JJ, Williams WM, Taylor NL (2006) Molecular phylogenetics of the clover genus (Trifolium – Leguminosae). Mol Phylogenet Evol 39:688–705

Gladis T, Hammer K (2003) Die *Brassica oleracea* Gruppe. Schriften des Vereins zur Erhaltung der Nutzpflanzenvielfalt 1, 79 pp. + Appendix. VEN, Lennestadt. German

Jones K (1962) Chromosomal status, gene exchange and evolution in *Dactylis*. Genetica 32:272–295

Lumaret R (1986) Doubled duplication of the structural gene for cytosolic phosphoglucose isomerase in the *Dactylis glomerata* L. polyploid complex. Mol Biol Evol 3:499–521

Lumaret R (1987) Differential degree in genetic divergence as a consequence of a long isolation in a diploid entity of *Dactylis glomerata* L. from the Guizhou region (China). Presentation to the 2nd symposium Paleoenvironment East Asia, Hong Kong, Jan 9–14, 1987

Lumaret R (1988) Cytology, genetics and evolution in the genus *Dactylis*. Crit Rev Plant Sci 7:55–91

Lumaret R, Barrientos E (1990) Phylogenetic relationships and gene flow between sympatric diploid and tetraploid plants of *Dactylis glomerata* (Gramineae). Plant Syst Evol 169:1615–6110

Lumaret R, Bowman CM, Dyer TA (1989) Autotetraploidy in *Dactylis glomerata* L.: further evidence from studies of chloroplast DNA variation. Theor Appl Genet 78:393–399

Lumerat R, Bretagnolle F, Maceira NO (1992) 2n gamete frequency and bilateral polyploidization in *Dactylis glomerata*. In: Mariana A, Tavoletti S (eds) Gametes with somatic chromosome number in the evolution and breeding of polyploid polysomic species: achievements and perspectives. Forage Plant Breeding Institute, Perugia, pp 15–21

Maly R, Hammer K, Lehmann Chr O (1987) Sammlung pflanzlicher genetischer Ressourcen in Süditalien—ein Reisebericht aus dem Jahre 1950 mit Bemerkungen zum Schicksal der Landsorten "in situ" und in der Genbank. Kulturpflanze 35:109–134 (German)

Sahuquillo E, Lumaret R (1999) Chloroplast DNA variation in *Dactylis glomerata* L. taxa endemic to the Macaronesian islands. Mol Ecol 8:1797–1803

Soltis DE, Soltis PS (1993) Molecular data and the dynamic nature of polyploidy. Crit Rev Plant Sci 12:243–273

Stebbins GL, Zohary D (1959) Cytogenetic and evolutionary studies in the genus *Dactylis*. I morphological, distribution and inter relationships of the diploid subspecies. Univ Calif Pub Bot 31:1–40

Stewart AV, Ellison NE (2010) Origins of diploid *Dactylis* subspecies based on nuclear ITS and chloroplast trnL sequences. Presentation to the molecular breeding of forage and turf conference, Buenos Aires 15th–19th March 2010

Stewart AV, Ellison N (2011) The genus *Dactylis*; wealth of wild species: role in plant genome elucidation and improvement. Springer, New York

Stewart AV, Joachimiak A, Ellison N (2009) Genomic and geographic origins of timothy (*Phleum* sp.) based on ITS and chloroplast sequences. In: Yamada T, Spangenberg G (eds) Molecular breedings of forage and turf. Springer, New York, pp 71–82

Stewart AV, Ellison NE, Martín Osorio VE (2013) Origins of diploid *Dactylis* from the Canary Islands as determined by DNA sequencing. In: Barth S, Milbourne D (eds) Breeding strategies for sustainable forage and turf grass improvement. Springer, Netherlands

Taberlet P, Gielly L, Pautou G, Bouvet, J (1991) Universal primers for amplification of three non-coding regions of chloroplast DNA. Plant Mol Biol 17:1105–1109

Williams WM, Ansari HA, Ellison NW, Hussain SW (2001) Evidence of three subspecies in *Trifolium nigrescens* Viv. Ann Bot Lond 87:683–691

Zurawski G, Clegg MT, Brown HD (1984) The nature of nucleotide sequence divergence between barley and maize chloroplast DNA. Genetics 106:735–749

Chapter 15
A Study on Cold Tolerance Transgenic Alfalfa (*Medicago sativa* L.) with the *AmDHN* Gene

Lizhen Nie, Hongkui Liu, Jie Sun, Heping ZHao, Jing Yi and Yongzhi Liu

Abbreviations

AmDHN	*Ammopiptanthus mongolicus* dehydrins gene
CAS	Cold acclimation specific
CBF	CRT binding factor
ProDH	Proline dehydrogenase
Pro	Proline
WT	Wild-type control (non-transgenic alfalfa)

15.1 Introduction

As low temperatures are a major environmental constraint limiting the growth, development, and distribution of many kinds of plants, the mechanisms underlying low temperatures injury have been the subject of frequent study. Low temperatures injury is usually caused by cellular dehydration, and the plasma membrane is the primary site of freezing injury (Thomashow 1999). However, plants employ multiple mechanisms to increase their tolerance to low temperatures, such as accumulation of compatible osmolytes (soluble sugars, glycine betaine, and proline (Pro)) and increased levels of cold-related genes.

A set of cold-induced proteins have also received particular attention. Among these, dehydrins (DHNs), also known as late embryogenesis abundant (LEA)

Y. Liu (✉) · L. Nie · J. Sun · H. ZHao · H. Liu · J. Yi
Inner Mongolia Academy of Agricultural & Animal Husbandry Sciences,
Hohhot 010031, P. R. China
e-mail: liuyongzhi@vip.163.com

H. Budak, G. Spangenberg (eds.), *Molecular Breeding of Forage and Turf*,
DOI 10.1007/978-3-319-08714-6_15

proteins have been evaluated. The accumulation of DHNs in plants may be induced by abscisic acid (ABA) or any environmental influence that causes dehydration of the cells, such as freezing or other low temperatures, heat, high salinity, or drought (Nylander et al. 2001; Wisniewski et al. 2006; Sun et al. 2013). It is proposed that DHNs can protect proteins and membranes from unfavorable structural changes caused by dehydration. Several studies have shown that the expression and accumulation of DHN play an important role in the acclimation of fruit trees to unfavorable temperatures. The expression of *CuCOR19*, a DHN detected in the leaves of *Citrus unshiu*, was found to be significantly upregulated by cold stress (4 °C) (Hara et al. 1999). Transgenic approaches have been used to illustrate the relationship between DHN accumulation and cold acclimation. Hara et al. found that over-expression of CuCOR19 could enhance cold tolerance in transgenic tobacco and prevent lipid peroxidation (Hara et al. 2003). Many DHNs have been shown to be associated with the regulation of cold tolerance in plants, rendering study of the function of the *Ammopiptanthus mongolicus* DHN highly relevant. However, information about the characteristics of this gene family in the *A. mongolicus* is limited. In the present study, both expressions of cold-related genes and the content of Pro increased significantly after 4 °C treatment in the transgenic plants, these results suggested that transgenic plants were more resistance to cold stress compared with the WT. These imply that *AmDHN* may regulate indirectly cold tolerance of the plants through regulating the expressions of cold-related genes and the content of Pro.

15.2 Methods

15.2.1 Construction of pBI-DHN Vector and Alfalfa Transformation

Medicago sativa (Zhongmu No. 2) seeds kindly provided by Dr. Linqing YU, Institute of Grassland Research, Chinese Academy of Agricultural Sciences, Hohhot, China) were used in this study. The coding sequence of *AmDHN* from *A. mongolicus* was PCR-amplified using primers 5′-CATGCTAGCGCTGCCTTTTA-3′ (forward) and 5′-AGCATGCATGTGCCTTTTTA-3′ (reverse) and digested by Sac I and Spe I. The fragment was inserted into the pBI121 vector by replacing multiple cloning sites. The resulting binary vector, pBI-DHN, was transferred into *Agrobacterium tumefaciens* strain LBA4404 using freezing/heat shock method. An alfalfa, Zhongmu No. 2, was used for agrobacterium-mediated transformation to generate transgenic plants (Nie et al. 2012). Vegetatively propagated plants from the Zhongmu No. 2 clone were used as wild-type (WT) control.

15.2.2 Growth and Treatments of M. Sativa

M. sativa (Zhongmu No. 2) seeds were soaked in 70% alcohol for approximately 2 min, thoroughly 10% sodium hypochlorite sterilizes 35 min and then thoroughly rinsed with sterile water. The seeds were transferred to petri plates at 25 °C in dark for 2 days until most of the seeds germinated. Seedlings were grown in incubator at 24/22 °C with 16/8 h photoperiod. When seedlings grew, approximately for 6 to 7 days, cotyledons and hypocotyls were cut with scissors, then putting into MS medium including 1 mg/L 2,4-D and 0.1 mg/L KT precultured for 3 days. Explants were used for agrobacterium-mediated transformation to generate transgenic plants (Nie et al. 2012). Well-established seedlings were transplanted to 4.5 in pots filled with vermiculite as soil. Cold treatment was performed by transferring plants to a 4 °C cold room, and leaves were collected in liquid nitrogen at different time points, stored at −80 °C.

15.2.3 Analysis of Gene Transcription

The gene expression pattern was studied using real-time quantitative PCR. Total RNAs were extracted from leaves of transgenic and control plants with Trizol reagent (Invitrogen) and treated with RNase-free DNase I (TaKaRa). The total RNAs were reverse transcribed into first-strand cDNA with SuperScript II reverse transcriptase (TakaRa), and the cDNAs obtained were used as templates for PCR amplification with specific primers. Gene-specific primers used for real-time quantitative RT-PCR were: 5′-CAT GCT AGC GCT GCC TTT TA-3′ and 5′-AGC ATGCATGTGCCTTTTTA-3′ for *AmDHN*; 5′-ACACTACTGGCTATGGAAG-3′ and 5′-CCT TCT TAC TCA AAA ACA G-3′ for *MsCAS17(L13415)*; 5′-GAAACTTAGATAAATGGGTG-3′ and 5′-GTA GCA GAG GTG GAA ATA G-3′ for *MsCBF2(EU139867)*; 5′-CAA CGA GGA GAT GCG ATT A-3 and 5′-ATGCCATTGCAGCAACAT-3′ for *MsCBF3(EU139868)*; 5′-ATC GCA ATC TCA AAG GTT-3′ and 5′-TGTTAATGGCTCTGGTCTC-3′ for *MsProDH*(AY556386.1). In addition, a housekeeping gene, *MtActin*, was used as control: 5′-ACG AGC GTT TCA GAT G-3′ and 5′-ACC TCC GAT CCA GAC A-3′ for *MsActin*. Previous studies have demonstrated that the use of actin as a reference to study the gene expression during cold treatment is reliable (Zhang et al. 2011a). Real-time PCR was performed in an optical 96-well plate with the real-time PCR system (Roche 480). Each reaction contained 7.5 μL 29 SYBR Green Master Mix reagents, 0.5 μL cDNA samples, and 0.6 μL 10 μM gene-specific primers in a final volume of 15 μL. The thermal cycle used was as follows: 95 °C for 2 min; 40 cycles of 95 °C for 30 s, 55 °C for 30 s, and 72 °C for 1 min. The expression values of the genes examined in the present study were normalized to the corresponding controls. We sequenced the PCR products and confirmed that the amplified sequences corresponded to their gene sequences.

15.2.4 Determination of Pro Content

The Pro content in leaves of transgenic and control alfalfa were determined as described by Bates et al. (Bates et al. 1973). Briefly, leaves were harvested, weighed, and extracted in 3 % sulfosalicylic acid. An aliquot of each extract (2 mL) was incubated with 2 mL of ninhydrin reagent (2.5 % [w/v] ninhydrin, 60 % [v/v] glacial acetic acid, and 40 % 6M phosphoric acid) and 2 mL of glacial acetic acid at 100 °C for 40 min, and the reaction was terminated in an ice bath. Toluene (4 mL) was added, followed by vortex and incubation at 23 °C for 24 h. The absorbance was measured at 520 nm.

15.2.5 Statistical Analysis

All data and figures were calculated and graphed by Microsoft Excel 2003.

15.3 Results

15.3.1 AmDHN Gene Expression Analysis

AmDHN gene was cloned from *A. mongolicus.* The ORF of *AmDHN* was placed under the control of the CaMV 35S promoter by inserting multiple cloning site of the binary vector pBI121. The resulting vector, pBI-DHN, was introduced into alfalfa by agrobacterium-mediated transformation and 100 independent transgenic plants were produced. PCR screening indicated that more than 60 % of the regenerated plants contained the target gene and Southern hybridization analysis revealed that 50 % of the plants contained the target gene, RT-PCR revealed that all plants expressed the transgene with various mRNA levels (data not shown). The expression pattern of the *AmDHN* gene in transgenic plants was detected by quantitative reverse transcription PCR (qRT-PCR) analysis, AmDHN transcript was quickly increased to a high level in different transgenic plants. However, lines five showed extremely high levels of AmDHN expression, expressions of the other lines were not the same (Fig. 15.1).

15.3.2 Effect of AmDHN Overexpression on Other Cold-Related Genes

qRT-PCR analysis was used to characterize the expression of a number of cold-related genes in the transgenic and nontransgenic alfalfa(WT). Total RNA was isolated

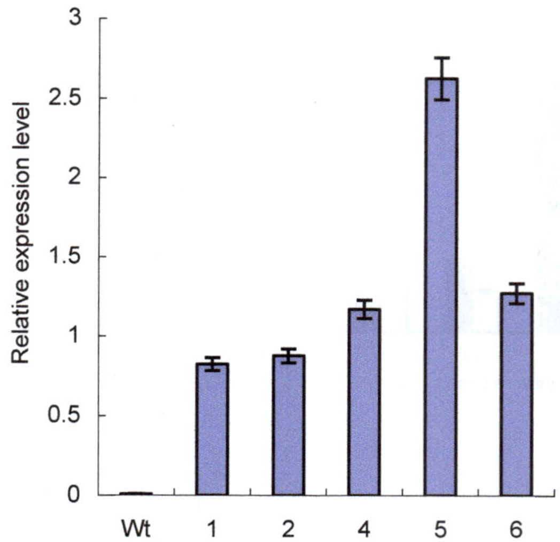

Fig 15.1 *AmDHN* expression level in the transgenic line as determined by qRT-PCR. *Wt* Wild-type plant. *1,2,4,5,6* Transgenic line

from leaves of the plants. Cold-related genes (*CBF2*, *BF3*, *ProDH*, and *CAS17*) were selected based on TBLASTN search against the TIGR *Medicago truncatula* gene index database (Quackenbush et al. 2000). After being treated with 4 °C, expressions of *CBF2*, *CBF3*, *ProDH*, and *CAS17* genes was 2.5, 4, 1.6, and 3 times higher than WT, respectively. Expressions of *CBF2* and *CBF3* increased significantly after being treated 5 h, reached an extremely significant level ($P<0.01$) respectively, as compared with WT plants. Similarly, expression of *CAS17* increased significantly after being treated 24 and 72 h, reached an extremely significant level respectively, as compared with WT. The level of *ProDH* transcripts increased dramatically 3 days after treatment by 4 °C, reached the maximum at 7th day($P<0.05$) (Fig. 15.2). These data indicate that during treatment at 4 °C, AmDHN mediates cold tolerance via a CBF pathway, consistent with a previous report (Zhang et al. 2011).

15.3.3 The Impact of Cold on the Content of Transgenic Alfalfa Pro

To test the transgenic alfalfa to low-temperature tolerance, Pro content was determined in transgenic alfalfa and WT. As shown in Fig. 15.3, the content of Pro in transgenic plants was significantly higher than WT at 7th day after the treatment, goes up to the highest at 14th day after the treatment ($P<0.01$). No significant differences at $P<0.05$ were detected at 3rd day after the treatment (Fig. 15.3). The analyses of Pro content in leaves showed that cold stress increased Pro content.

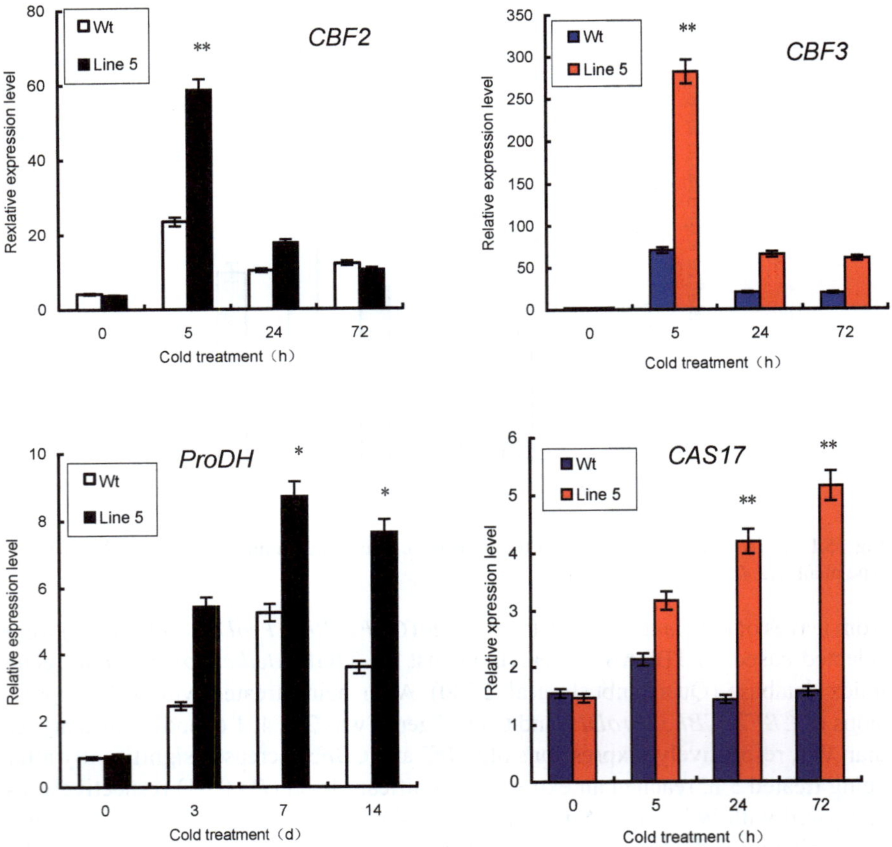

Fig. 15.2 Expression profile analysis of stress-relative gene in the transgenic plant in response to cold stress. ** mean extremely significant difference ($P<0.01$). * mean significant difference ($P<0.05$)

15.4 Discussion

Dehydrins (DHNs), belonging to LEA D11 family, are among the most commonly observed proteins induced by environmental stress associated with dehydration or low temperature. LEA is known typically to accumulate during seed development and is thought to play a role in protecting the embryo during desiccation. Studies with soybean indicated that accumulation of LEA proteins during embryogenesis might reduce the extent of desiccation-induced electrolyte leakage in immature seeds, suggesting a role in membrane protection (Dure et al. 1981). So we speculated that dehydration gene involved in stress resistance of plants. Previous studies showed that AmDHN was induced by drought, osmotic stress and ABA (Sun et al. 2013). Our findings found that expressions of *CBF2*, *CBF3*, *CAS17*, and *ProDH* in transgenic alfalfa increased significantly after being treated with 4 °C, as compared

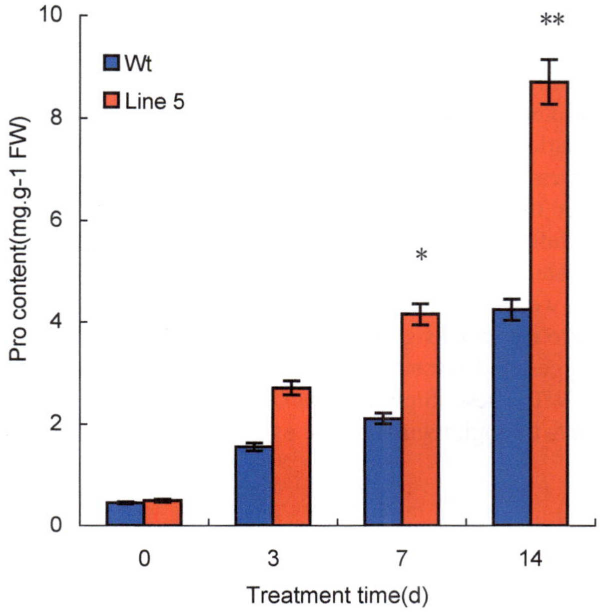

Fig. 15.3 Proline (Pro) content in the transgenic plants in response to cold stress. *Wt* Wild-type plant. Line *5* transgenic plant. ** mean extremely significant difference ($P<0.01$). * mean significant difference ($P<0.05$)

with WT. Under normal conditions, expressions of cold-related genes in transgenic plant and WT are similar; however, all genes are upregulated dramatically after being treated with 4 °C in transgenic plant. These results are consistent with previous studies (Hara et al. 2003). These results demonstrate that accumulation of higher CBF and CAS transcript levels in transgenic alfalfa may play a role in conferring greater tolerance of transgenic alfalfa to cold than that of nontransgenic alfalfa (Zhang et al. 2010). The CBF cold response pathway plays an important role in cold tolerance in many plant species (Thomashow 2010). In *Arabidopsis*, cold acclimation-induced freezing tolerance is positively correlated with expression levels of CBF1-3 between ecotypes of Ws-2 and Cvi-1 (Cook et al. 2004). For instance, overexpression of AtCBF1 and AtCBF3 enhanced freezing tolerance in nonacclimated plants and induced soluble sugars and Pro accumulated in plants under warm conditions. Therefore, a complex network among CBF1, CBF2, and CBF3 may exist in the regulation of freezing tolerance in plants. Our data showed that expressions of MsCBF2 and MsCBF3 were higher than that of nontransgenic plant in response to cold treatment (Novillo et al. 2007; Zhang et al. 2011b). We speculate that CBF2 and CBF3 may have a greater effect on cold tolerance in transgenic plant.

There are many CAS genes found in alfalfa, and their expression is positively correlated with freezing tolerance (Monroy and Dhindsa 1995). In the present study, we found that the transcript of *MsCAS17* in transgenic plant was accumulated at higher levels than nontransgenic plant during cold treatment (Fig. 15.2). It is conceivable that the higher levels of *MsCAS17* may lead to greater amounts of dehydrin proteins, thus protecting transgenic plants from dehydration under cold stress.

The accumulation of Pro is a common phenomenon in plants exposed to a number of abiotic stresses, including cold, drought, and salinity (Zhao et al. 2009). Our findings revealed that Pro began to increase after 7th day at 4 °C and Pro contents in transgenic alfalfa were higher than in nontransgenic alfalfa (Fig. 15.3). These results are consistent with previous studies.

In summary, we demonstrated that both expressions of cold-related genes and the content of Pro increased significantly after the treatment in the transgenic plant, these results suggested that transgenic plant was more resistance to cold stress compared with the WT. These imply that *AmDHN* may regulate indirectly cold tolerance of the plants through regulating the expressions of cold-related genes and the content of Pro.

References

Bates LS, Waldren RP, Teare ID (1973) Rapid determination of free proline for water-stress studies. Plant Soil 39:205–207

Cook D, Fowler S, Fiehn O, Thomashow MF (2004) A prominent role for the CBF cold response pathway in configuring the low temperature metabolome of *Arabidopsis*. Proc Natl Acad Sci 101:15243–15248

Dure L 3RD, Greenway SC, Galau GA (1981) Developmental biochemistry of cottonseed embryogenesis and germination: changing messenger ribonucleic acid populations as shown by in vitro and in vivo protein synthesis. Biochemistry 20(14):4162–4168

Hara M, Wakasugi Y, Ikoma Y, Yano M, Ogawa K, Kuboi T (1999) cDNA sequence and expression of a cold-responsive gene in *Citrus unshiu*. Biosci Biotechnol Biochem 63(2):433–437

Hara M, Terashima S, Fukaya T, Kuboi T (2003) Enhancement of cold tolerance and inhibition of lipid peroxidation by citrus dehydrin in transgenic tobacco. Planta 217:290–298

Monroy AF, Dhindsa RS (1995) Low temperature signal transduction:induction of cold acclimation-specific genes of alfalfa by calcium at 25 °C. Plant Cell 7:321–331

Nie LZ, Guo JF, Sun J, Sun GQ, Liu YZ (2012) Research on transformation of DHN Gene from *Ammopiptanthus mongolicus* into alfalfa (*Medicago sativa L.*). Acta Agric Boreali-Sinica 27(3):96–101

Novillo F, Medina J, Salinas J (2007) *Arabidopsis* CBF1 and CBF3 have a different function than CBF2 in cold acclimation and define different gene classes in the CBF regulon. Proc Natl Acad Sci U S A 104:21002–1007

Nylander M, Svensson J, Palva ET, Welin BV (2001) Stress-induced accumulation and tissue-specific localization of dehydrins in *Arabidopsis thaliana*. Plant Mol Biol 45:263–279

Quackenbush J, Liang F, Holt I, Pertea G, Upton J (2000) The TIGR Gene Indices: reconstruction and representation of expressed gene sequences. Nucleic Acids Res 28:141–145

Sun J, Nie LZ, Sun GQ, Guo JF, Liu YZ (2013) Cloning and characterization of dehydrin gene from *Ammopiptanthus mongolicus*. Mol Biol Rep 40(3):2281–2291

Thomashow MF (1999) Plant cold acclimation: freezing tolerance genes and regulatory mechanisms. Annu Rev Plant Physiol Plant Mol Biol 50:571–599

Thomashow MF (2010) Molecular basis of plant cold acclimation: insights gained from studying the CBF cold response pathway. Plant Physiol 154:571–577

Wisniewski ME, Bassett CL, Renaut J, Farrell RJ, Tworkoski T, Artlip TS (2006) Differential regulation of two dehydrin genes from peach (*Prunus persica*) by photoperiod, low temperature and water deficit. Tree Physiol 26(5):575–584

Zhang H, Mao X, Wang C, Jing R (2010) Overexpression of a common wheat gene TaSnRK2.8 enhances tolerance to drought, salt and low temperature in *Arabidopsis*. PLoS ONE 5(12):e16041

Zhang LL, Zhao MG, Tian QY, Zhang WH (2011a) Comparative studies on tolerance of *Medicago truncatula* and *Medicago falcata* to freezing. Planta 234:445–457

Zhang YJ, Yang JS, Guo SJ, Meng JJ, Zhang YL, Wan SB, et al (2011b) Over-expression of the *Arabidopsis* CBF1 gene improves resistance of tomato leaves to low temperature under low irradiance. Plant Biol 13:362–367

Zhao MG, Chen L, Zhang LL, Zhang WH (2009) Nitric reductase-dependent nitric oxide production is involved in cold acclimation and freezing tolerance in *Arabidopsis*. Plant Physiol 151:755–767

Chapter 16
The Intron in an Albumin Gene from Sunflower Increases Expression of *SFA8*

Zhang Qiong-Lin, Heping ZHao, Lizhen Nie, Jing Yi and Jie Sun

16.1 Introduction

It is very important of the protein content in plants, especially in forage plants. To improve forage protein content is an important job for many breeders. Traditional breeding methods are mainly focused on cross-breeding, recurrent selection methods. However, these methods cost a longer period and a lot of work. With advances in genetic engineering technology, more and more transgenic technology has been applied to breeding works, greatly reducing breeding cycle, improving work efficiency.

Sunflower is a very important oil crop in the world (Ren et al. 1997) . Sunflower is rich in albumin, and 2S albumin plays important roles in resistant to degradation of rumen bacteria, and has potential for improving the nutritional value of pasture plant by genetic engineering. In 1993, Tabe inserted the sunflower albumin gene into alfalfa and other plants to improve the nutritional value of forage and increased the protein content of alfalfa (Tabe et al. 1993) , but the expression of exogenous protein in transgenic plants is need to be improved. The intron editing process in higher plant is an important part of eukaryotic expression; particularly alternative splice can regulate gene expression. The intron in higher plant gene plays important roles in the regulation of gene expression, such as changing gene expression patterns, enhancing the level of gene expression and activation of gene expression (Xie and Wu 2002). The traditional transgenic methods are used without any intron, because the first strand cDNA was used as the template when the PCR amplification

J. Sun (✉) · H. ZHao · L. Nie · Y. Jing · Z. Qiong-Lin
Inner Mongolia Academy of Agricultural & Animal Husbandry Sciences,
Hohhot 010031, P. R. China
e-mail: sunjienmg@126.com

H. Budak, G. Spangenberg (eds.), *Molecular Breeding of Forage and Turf*,
DOI 10.1007/978-3-319-08714-6_16

of the coding region of the gene sequence. The plant introns can not simply be seen as useless DNA sequence and a large number of experiments showed that plant intron is not only functional, but also plays an important role in the process of regulation of gene expression. The first intron of Sh1 gene from maize plays a very important role in enhancing gene expression (Clancy and Hannah 2002).

Although there have been experiments on obtaining transgenic plants with albumin, the expression level of albumin gene is not very high. In this experiment, we cloned gDNA and cDNA sequences of sunflower albumin and successfully transformed into Arabidopsis genome. Quantitative PCR analysis showed that SFA8 expression levels in transgenic Arabidopsis with intron were obviously higher than those in transgenic Arabidopsis without intron. These results showed that the intron from sunflower can improve the expression levels of *SFA8* gene in transgenic Arabidopsis, which provides a theoretical basis in use of sunflower *SFA8* gene.

16.2 Materials and Methods

16.2.1 Experiment Materials

Sunflower and *Arabidopsis thaliana* seeds were obtained from Inner Mongolia Academy of Agricultural & Animal Husbandry Sciences. Agrobacterium GV3101 and plant expression vector pBI121 are stored in our laboratory.

16.2.2 Primers Designation

Based on the sequence of SFA8 gene, one pair of primers were designed to construct plant expression vector:

SFA8f: 5′-ggtctagaATGGCAAGGTTTTCGATCGTGTTTG-3′

SFA8r: 5′-ccggatccCATTTGGCATGGTTGGGACATGAG-3′

The *Xba* I and *Bam*H I enzyme sites were underlined.

16.2.3 Cloning of cDNA and gDNA of SFA8

Total RNA was extracted from sunflower seeds by a guanidine thiocyanate extract method. Any contaminated genomic DNA was removed by incubating the total RNA with RNase-free DNase (Promega) at 37 °C for 30 min. And total RNA was used to synthesize the first-strand cDNA. The first-strand cDNA was used as the temple to clone cDNA of SFA8 gene with primers, SFA8f and SFA8r. Thermocycling was performed at 35 cycles with 94 °C for 30 s, 56 °C for 30 s and 72 °C for 30 s, and an additional polymerization step at 72 °C for 5 min. The genomic DNA was isolated from leaves of sunflower using CTAB method. The genomic DNA was used as the temple to clone gDNA of SFA8 gene with primers, SFA8f and SFA8r. Thermocycling was performed at 35 cycles with 94 °C for 30 s, 56 °C for 30 s and 72 °C for 1 min, and an additional polymerization step at 72 °C for 5 min. PCR products were separated by electrophoresis on a 1 % agarose gel stained with ethidium bromide, purified using the DNA gel extraction kit (TaKaRa, Japan). The products were cloned into the pMD-18 T (TaKaRa, Japan) vector and then transformed into *E. coli* DH5*a*. And recombinant plasmids were sequenced by Sangon Biotech Co., Ltd (Shanghai, China).

16.2.4 Vector Construction

The plasmids, pMD-SFA8 g, pMD-SFA8c and pBI121 were digested with XbaI and BamHI enzymes. The DNA fragments were separated by electrophoresis on a 1 % agarose gel, purified using the DNA gel extraction kit. The cDNA and gDNA fragments were inserted into the same enzyme sites in pBI121 to generate plant expression vector. The DNA recombinations were characterized by being digested with XbaI and BamHI enzymes.

16.2.5 Arabidopsis Transformation

The plant expression vectors, $35S_{pro}$::SFA8c and $35S_{pro}$::SFA8 g were extracted and transformed into Agrobacterium GV3101. The transformed Agrobacterium were identified by PCR methods. To generate transgenic Arabidopsis plants, Agrobacterium GV3101 transformed with plasmid $35S_{pro}$::SFA8c or $35S_{pro}$::SFA8 g was used to infect Arabidopsis using the floral dip method (Chao et al. 2009). Transformed Arabidopsis seeds were selected on ½-strength MS medium containing 50 $mg{\cdot}L^{-1}$ Kanamycin and 0.8 % agar. After 15 days, seedlings keeping green were selected and transferred into soil.

Identification and expression analysis of transgenic plants

The genomic DNA of transgenic Arabidopsis was extracted by CTAB method. To identify transgenic plants, genomic DNA was used as the temple, with pBI121 vector primers, pBI-f and pBI-r. Thermocycling was performed at 35 cycles with 94 °C for 30 s, 55 °C for 30 s and 72 °C for 1 min, and an additional polymerization step at 72 °C for 5 min.

The total RNA was extracted from each transgenic Arabidopsis (T_2 line). Any contaminated genomic DNA was removed by incubating the total RNA with RNase-free DNase. And total RNA was used to synthesize the first-strand cDNA with the random primer. Based on the SFA8 gene sequence, one pair of primers, SFA8-RT-f (5′-TTTAGGAGAAAGGAGGCAAGTTAG-3′) and SFA8-RT-r (5′-TTTAGGAGAAAGGAGGCAAGTTAG-3′), was designed to analyze the expression level of SFA8 gene in transgenic plants by qRT-PCR. *Actin* was used as an internal

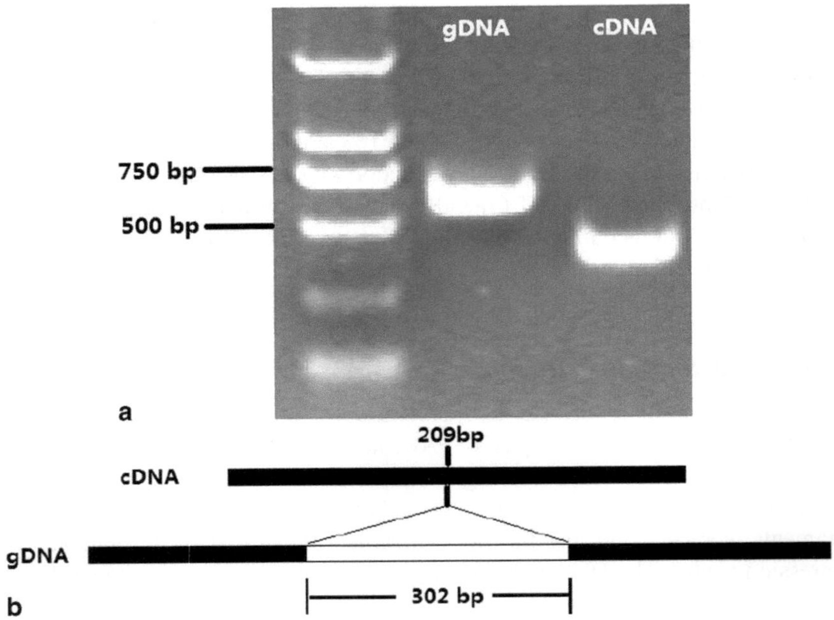

Fig. 16.1 cDNA and gDNA of *SFA8* gene **a** PCR amplification of cDNA and gDNA of *SFA8* gene; **b** Structure of cDNA and gDNA of *SFA8* gene

control. The two-step thermal cycling profile used was 15 s at 95 °C and 1 min at 68 °C. The qRT-PCR reactions were performed in biological triplicates using total RNA samples extracted from three independent plant materials. The comparative ΔΔthreshold cycle method (Chao et al. 2009) was used to evaluate the relative quantities of each amplified product in the samples.

16.3 Results

16.3.1 Cloning of SFA8 cDNA and gDNA

The full-length cDNA was obtained by PCR with SFA8f and SFA8r as primers and cDNA as the temple. PCR product was analyzed on 1 % agrose gel and result showed that a 500-bp fragment was cloned (Fig. 16.1). This fragment was inserted into clone vector pMD18-T and the resulting plasmid was transformed into *E. coli* DH5 competent cells. PCR analysis was carried on and the results showed that the positive clones were selected for sequencing. Analysis of sequence balst showed that there was no base-mutant or frame-shift mutation in clones sequenced. The result showed that the cDNA sequence of *SFA8* gene was cloned successfully, which was named as pMD-SFA8c. The gDNA was obtained by PCR with SFA8f

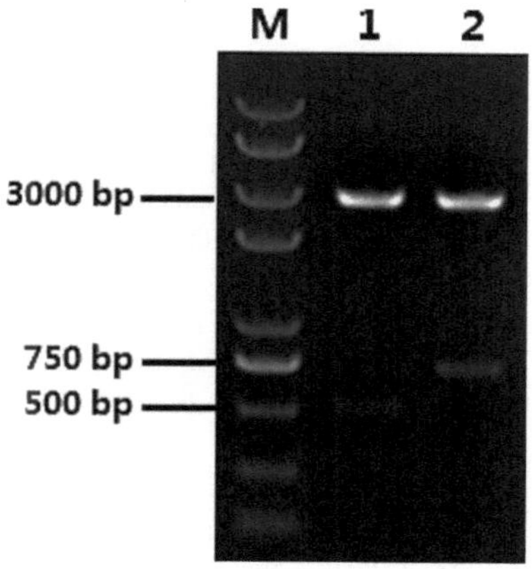

Fig. 16.2 Restrict Enzymes analysis of clone vectors. M: DNA Marker DL2000 Plus II; 1: pMD-SFA8c digested with restrict enzymes; 1: pMD-SFA8 g digested with restrict enzymes

and SFA8r as primers and genomic DNA as the temple. PCR product was analyzed on 1 % agrose gel and result showed that a 750-bp fragment was cloned (Fig. 16.1). This fragment was inserted into clone vector pMD18-T and the resulting plasmid was transformed into *E. coli* DH5 competent cells. PCR analysis was carried on and the results showed that the positive clones were selected for sequencing. Analysis of sequence balst showed that there existed one 302-bp intron in the SFA8 gDNA sequence. The result showed that the gDNA sequence of *SFA8* gene was cloned successfully, which was named as pMD-SFA8 g.

16.3.2 Vector Construction

Plasmids, pMD-SFA8c, pMD-SFA8 g and pBI121, were digested with restrict enzymes, XbaI and BamHI. The digested products were analyzed on 1 % agrose gel and results showed that there existed one about 500-bp DNA fragment in the digested pMD-SFA8c (Fig. 16.2) and one about 750-bp DNA fragment in the digested pMD-SFA8 g (Fig. 16.2). The target fragments were purified, which were used to be ligased by T4 DNA ligase. The products were transformed into *E. coli* DH5 competent cells. PCR analysis was carried on and the positive clones were selected for plasmid extract. Enzyme analysis of the two constructed plasmids showed that there was one DNA fragment in each digested products on 1 % agrose gel (Fig. 16.3), which was consistent with prediction. The two constructed plant expression vectors were named as $35S_{pro}$::SFA8c and $35S_{pro}$::SFA8 g respectively.

16.3.3 PCR Analysis of Transgenic Arabidopsis

Transgenic Arabidopsis seeds were selected on the 1/2MS medium with 50 mg·L^{-1} Kanamycin. Eight resistant seedlings were selected by the antibiotic. All of them

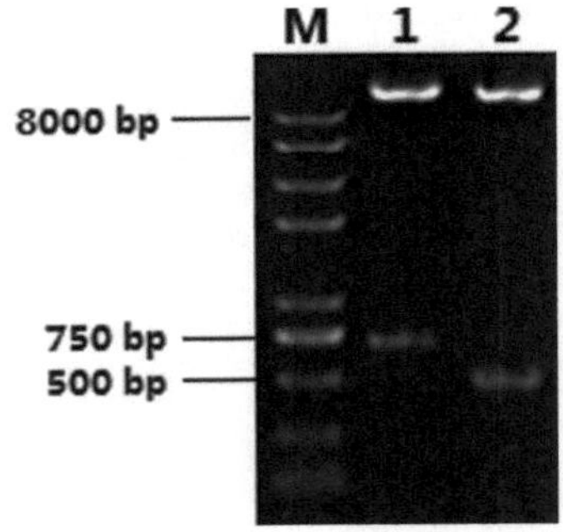

Fig. 16.3 Restrict Enzymes analysis of plant expression vectors. M: DNA Marker DL2000 Plus II; 1: 35_{pro}::SFA8 g digested with restrict enzymes; 1: 35_{pro}::SFA8c digested with restrict enzymes

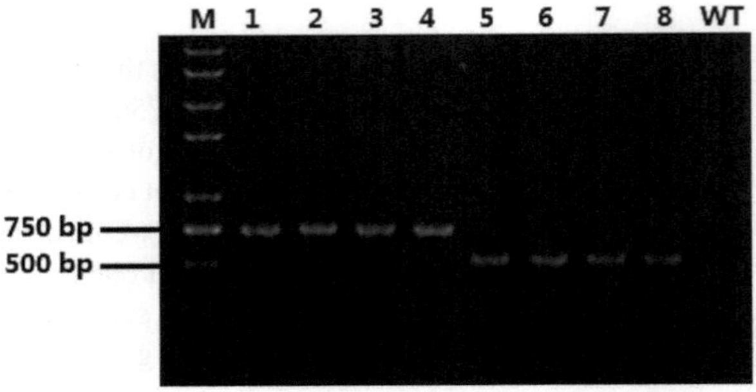

Fig. 16. 4 PCR analysis of transgenic Arabidopsis. M: DNA Marker DL2000 Plus II; 1-4:transgenic Arabidopsis with 35_{pro}::SFA8 g; 5-8: transgenic Arabidopsis with 35_{pro}::SFA8 g

were used for PCR analysis. PCR was carried on with SFA8f and SFA8r as primers and genomic DNA from transgenic Arabidopsis as the temple. PCR products were analyzed on 1 % agrose gel. The results showed that one fragment of predicted size could be cloned by PCR from transgenic plants, but as the negative control, wild type Arabidopsis could not be amplified any DNA fragment (Fig. 16.4). The results displayed that the cDNA and gDNA of SFA8 gene had been inserted into genomic DNA of transgenic Arabidopsis.

16.3.4 Expression analysis of SFA8 gene in transgenic Arabidopsis

The total RNA was extracted from each transgenic Arabidopsis (T_2 line). Any contaminated genomic DNA was removed by incubating the total RNA with RNase-free DNase. And total RNA was used to synthesize the first-strand cDNA. The SFA8 gene expression levels were analyzed by qRT-PCR with primers, SFA8-RT-f and SFA8-RT-r. The results showed that SFA8 gene expression levels could be detected in all 8 transgenic Arabidopsis lines and expression levels in transgenic plants with $35S_{pro}$::SFA8 g were obviously higher than those with $35S_{pro}$::SFA8c (Fig. 16.5). The results displayed that the SFA8 intron could obviously improve the expression levels of SFA8 gene in transgenic Arabidopsis.

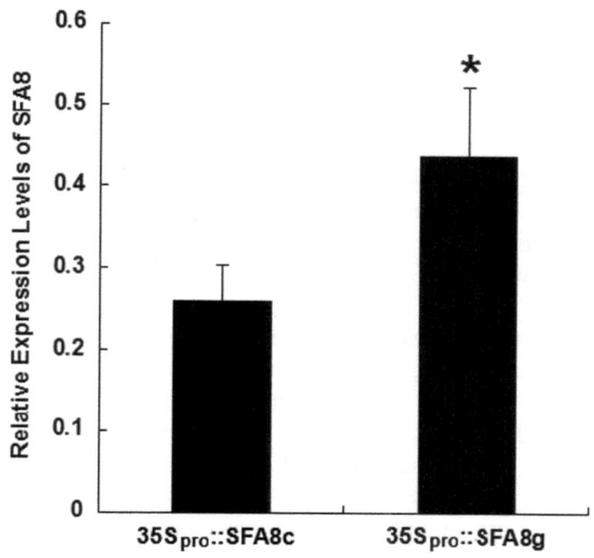

Fig. 16.5 Expression analysis of SFA8 gene in transgenic Arabidopsis plants

16.4 Discussion

By transgene method, expression of genes related with good traits has been reported in many other species. As a new technique, molecular biology has been developed for only a few decades, but has made remarkable achievements in a short time. In 1985, Stalker, transferred a bacterial gene ESPS to tobacco, successfully obtaining glyphosate-resistant plants (Stalker et al. 1985). In 2010, by transgenic technology, alfalfa zinc finger protein gene *MsZFN* was inserted into tobacco and was successfully expressed (Chao et al. 2010). In 2013, by transgenic technology, alfalfa *MsFRI-L* gene was transferred into Arabidopsis genomes, causing late flowering time in transgenic plants (Chao et al. 2013). In this experiment, the sunflower albumin *SFA8* gene was transformed into *Arabidopsis thaliana* plants attempting overexpression of the sunflower albumin in transgenic plants. Improving of albumin by overexpression of the SFA8 gene in transgenic plants has been previously reported. In 1991, Schroeder etc. cloned chicken albumin and successfully transferred it into alfalfa (Schroeder et al. 1991). In 1993, Tabe etc. transferred the sunflower albumin gene into alfalfa and other plants to improve the nutritional value of forage (Tabe et al. 1993). In 2009, Zhang etc. inserted successfully pea albumin PA1 gene into alfalfa, and they found that expression of PA1 gene in transgenic alfalfa increased methionine and cysteine from 0.1 to 0.4% (Zhang and Jia 2009). When designing the experiment, we did not follow the traditional method, considering that the sunflower albumin intron may play a regulatory role in albumin gene expression, so two plant expression vectors were constructed, one of which contains SFA8 complete coding region, and the other containing the coding region with the intron. In eukaryotes, the intron plays an important role in regulating the gene expression, which is deemed as an important element in improving expression of exogenous genes in transgenic plants. A large number of experiments show that the presence of the intron can improve the expression levels of exogenous genes in transgenic plants. In 1995, Donath etc. inserted the first intron of maize sh1 gene upstream of the cat gene, finding that the expression level of *cat* was increased by 60–150 times in transgenic plants (Donath et al. 1995). The first intron of rice EP-SPS gene can increase the expression level of the GUS gene, six times in maximum (Xu et al. 2003). Arabidopsis H2A intron can increase GUS expression in the transgenic plants (Zou et al. 2008). By real-time quantitative RT-PCR method, the two transgenic plant expression analyses showed that expression levels of SFA8 gene in transgenic plants containing SFA8 intron were much higher than those without intron. These results indicated that SFA8 intron can enhance the expression of the SFA8 gene in Arabidopsis

In this study, we isolated sunflower SFA8 cDNA and gDNA sequences, both including a coding region of SFA8 gene. A 302-bp of intron is in the gDNA sequence. Two plant expression vectors were constructed and transgenic plants were successfully obtained. Confirmed by qRT-PCR analysis, SFA8 intron can increase the expression of *SFA8* gene in Arabidopsis. These findings would provide helps in improving plants and increasing plant protein contents.

References

Chao YH, Kang JM, Sun Y, Yang QC, Wang PQ, Wu MS, Li Y, Long RC, Qin ZH (2009) Molecular cloning and characterization of a novel gene encoding zinc finger protein from *Medicago sativa* L. Mol Biol Reports 36(8):2315–2321

Chao YH, Kang JM, Wang PQ, Yang QC (2010) Cloning of MsZFN gene from alfalfa and transformation into tobacco. Chin J Grassl 32(6):21–26

Chao YH, Yang QC, Kang JM, Zhang TJ, Sun Y (2013) Expression of the alfalfa FRIGIDA-Like gene, MsFRI-L delays flowering time in transgenic *Arabidopsis thaliana*. Mol Biol Reports 40(3):2083–2090

Clancy M, Hannah LC (2002) Splicing of the maize Sh1 first intron is essential for enhancement of gene expression, and a T-rich motif increases expression without affecting splicing. Plant Physiol 130(2):918–929

Donath M, Mendel R, Cerff R, Martin W (1995) Intron-dependent transient expression of the maize GapA1 gene. Plant Mol Biol 28(4):667–676

Ren J, Zheng XQ, Liu XL, Lin GP, Xia WS (1997) Preparation and properties of 2S globular proteins in the sunflower meal. Cereals Oils Process 1997(7):81–83

Schroeder HE, Khan MRI, Knibb WR, Spencer D, Higgins TJV (1991) Expression of a chicken ovalbumin gene in three lucerne cultivars. Funct Plant Biol 18(5):495–505

Stalker DM, Hiatt WR, Comai L (1985) A single amino acid substitution in the enzyme 5-enolpyruvylshikimate-3-phosphate synthase confers resistance to the herbicide glyphosate. J Biol Chem 260(8):4724–4728

Tabe LM, Higgins CM, McNabb WC, Higgins TJV (1993) Genetic engineering of grain and pasture legumes for improved nutritive value. Genetica 90(2–3):181–200

Xie XZ, Wu NH (2002) The intron of higher plant gene. Chinese Sc Bull 47(10):731–737

Xu JW, Feng DJ, Song GS, Wei XL, Chen L, Wu XL, Li XG, Zhu Z (2003) The first intron from rice EPSP synthetase increases expression of exogenous gene. Sci China (Ser. C) 33(3):224–230

Zhang GN, Jia JF (2009) Cloning of PA1 gene from *Pisum sativum* and genetic transformation on *Medicago sativa*. Acta Prataculturae Sin 18(3):117–125

Zou Z, Wu G, Xiao L, Wu YH, Nie SJ, Xiao N, Lu CM (2008) Effects of botanical intron on expression pattern of GUS gene. Biotechnol Bull 2008(6):78–82

Index

H. Budak, G. Spangenberg (eds.), *Molecular Breeding of Forage and Turf*,
DOI 10.1007/978-3-319-08714-6

MIX
Papier aus verantwortungsvollen Quellen
Paper from responsible sources
FSC® C105338

If you have any concerns about our products,
you can contact us on
ProductSafety@springernature.com

In case Publisher is established outside the EU,
the EU authorized representative is:
Springer Nature Customer Service Center GmbH
Europaplatz 3, 69115 Heidelberg, Germany

Printed by Libri Plureos GmbH
in Hamburg, Germany